T0325706

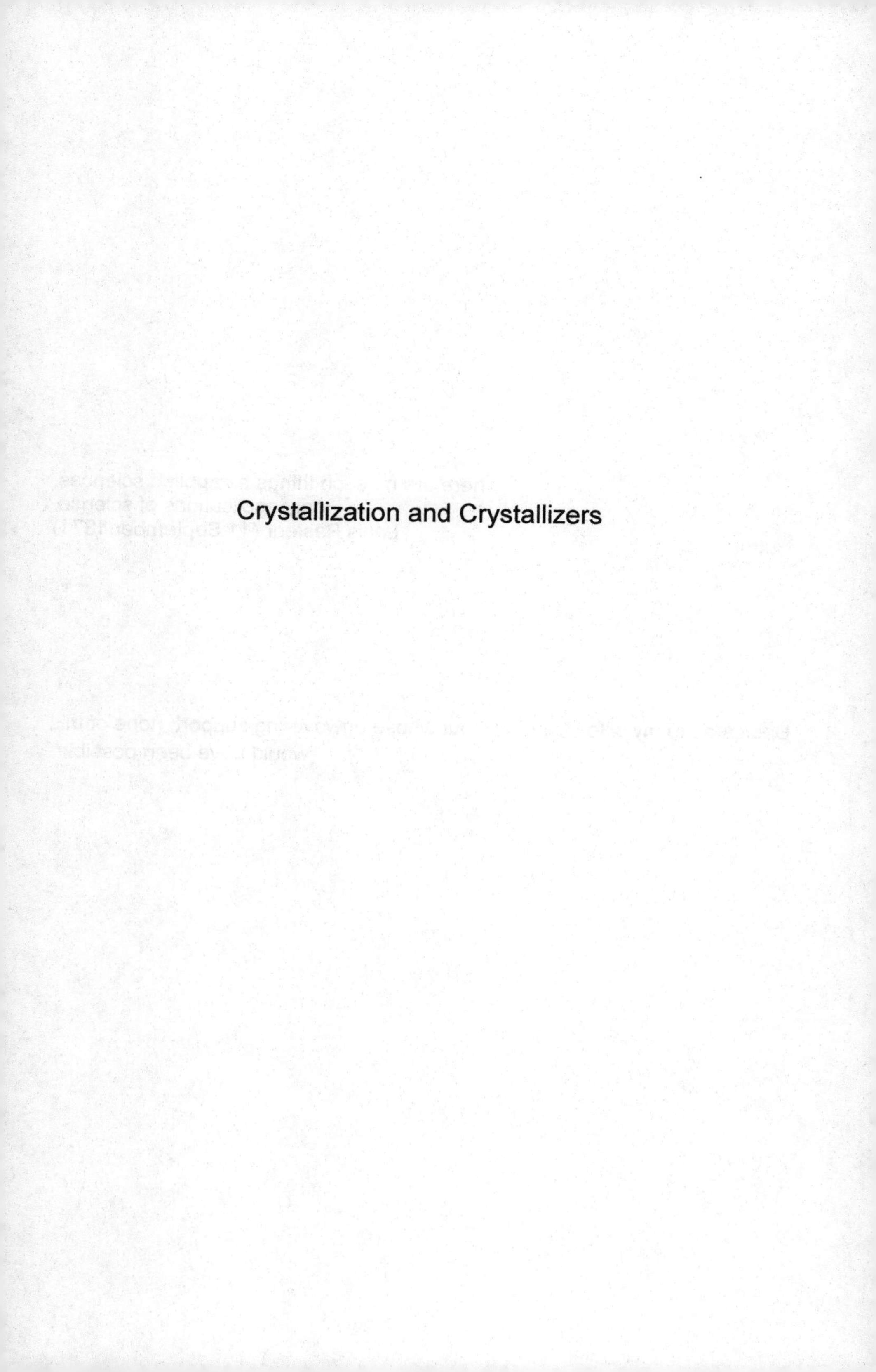

Crystallization and Crystallizers

There are no such things as applied sciences,
only applications of science.
Louis Pasteur (11 September 1871)

Dedicated to my wife, Anne, without whose unwavering support, none of this would have been possible.

Industrial Equipment for Chemical Engineering Set
coordinated by
Jean-Paul Duroudier

Crystallization and Crystallizers

Jean-Paul Duroudier

First published 2016 in Great Britain and the United States by ISTE Press Ltd and Elsevier Ltd

ISTE Press Ltd
27-37 St George's Road
London SW19 4EU
UK

www.iste.co.uk

Elsevier Ltd
The Boulevard, Langford Lane
Kidlington, Oxford, OX5 1GB
UK

www.elsevier.com

British Library Cataloguing-in-Publication Data
A CIP record for this book is available from the British Library
Library of Congress Cataloging in Publication Data
A catalog record for this book is available from the Library of Congress
ISBN 978-1-78548-186-4

Printed and bound in the UK and US

Contents

Preface

The observation is often made that, in creating a chemical installation, the time spent on the recipient where the reaction takes place (the reactor) accounts for no more than 5% of the total time spent on the project. This series of books deals with the remaining 95% (with the exception of oil-fired furnaces).

It is conceivable that humans will never understand all the truths of the world. What is certain, though, is that we can and indeed must understand what we and other humans have done and created, and, in particular, the tools we have designed.

Even two thousand years ago, the saying existed: "faber fit fabricando", which, loosely translated, means: "*c'est en forgeant que l'on devient forgeron*" (a popular French adage: *one becomes a smith by smithing*), or, still more freely translated into English, "practice makes perfect". The "artisan" (faber) of the 21st Century is really the engineer who devises or describes models of thought. It is precisely that which this series of books investigates, the author having long combined industrial practice and reflection about world research.

Scientific and technical research in the 20th century was characterized by a veritable explosion of results. Undeniably, some of the techniques discussed herein date back a very long way (for instance, the mixture of water and ethanol has been being distilled for over a millennium). Today, though, computers are needed to simulate the operation of the atmospheric distillation column of an oil refinery. The laws used may be simple statistical correlations but, sometimes, simple reasoning is enough to account for a phenomenon.

Since our very beginnings on this planet, humans have had to deal with the four primordial "elements" as they were known in the ancient world: earth, water, air and fire (and a fifth: aether). Today, we speak of gases, liquids, minerals and vegetables, and finally energy.

The unit operation expressing the behavior of matter are described in thirteen volumes.

It would be pointless, as popular wisdom has it, to try to "reinvent the wheel" – i.e. go through prior results. Indeed, we well know that all human reflection is based on memory, and it has been said for centuries that every generation is standing on the shoulders of the previous one.

Therefore, exploiting numerous references taken from all over the world, this series of books describes the operation, the advantages, the drawbacks and, especially, the choices needing to be made for the various pieces of equipment used in tens of elementary operations in industry. It presents simple calculations but also sophisticated logics which will help businesses avoid lengthy and costly testing and trial-and-error.

Herein, readers will find the methods needed for the understanding the machinery, even if, sometimes, we must not shy away from complicated calculations. Fortunately, engineers are trained in computer science, and highly-accurate machines are available on the market, which enables the operator or designer to, themselves, build the programs they need. Indeed, we have to be careful in using commercial programs with obscure internal logic which are not necessarily well suited to the problem at hand.

The copies of all the publications used in this book were provided by the *Institut National d'Information Scientifique et Technique* at Vandœuvre-lès-Nancy.

The books published in France can be consulted at the *Bibliothèque Nationale de France*; those from elsewhere are available at the British Library in London.

In the in-chapter bibliographies, the name of the author is specified so as to give each researcher his/her due. By consulting these works, readers may gain more in-depth knowledge about each subject if he/she so desires. In a reflection of today's multilingual world, the references to which this series points are in German, French and English.

The problems of optimization of costs have not been touched upon. However, when armed with a good knowledge of the devices' operating parameters, there is no problem with using the method of steepest descent so as to minimize the sum of the investment and operating expenditure.

1

Various Properties of Crystals

1.1. Structure

1.1.1. *Introduction: crystallization energy*

The structure of a crystal describes its internal configuration, that is, the disposition of atoms and molecules in the periodic lattice. As far as this is concerned, we can refer to the works of both Rousseau [ROU 99] and Novick [NOV 95]. At present, we will focus on that which is known as crystallization in a vacuum. Applied to a molecule or an atom, this energy is the reduction of potential energy initially taken at infinity and then integrated into the crystal.

The principle of this calculation method consists of accepting that the interaction of two molecules is the total of the atomic interactions of the combined molecules taken two by two. Thus, the interaction potential of molecule k with molecule ℓ is

$$U_{k\ell} = \frac{1}{2}\sum_{i}\sum_{j} V_{ij}$$

i: current index of molecule k atoms

j: current index of molecule ℓ atoms

Coefficient 1/2 avoids the need to count the same potential twice. This potential is the total of three terms:

– the electrical term;

– the dispersion term (Van der Waals);

– hydrogen bond.

1) The electrical term:

This is due to the charges of atomic nuclei corrected with the masking effect due to the electron cloud that surrounds atoms in a molecule. This effect is often announced by a reduction of atomic (nucleus) charge in the order of 30%. The electron interaction potential [VES 99] is as follows:

$$V_{elec} = \frac{Ze}{4\pi\varepsilon_o r} - \int_{\Omega} \frac{\rho(r')dr'}{4\pi\varepsilon_o (r' - r)}$$

r: distance separating nucleus from the point at which we assess potential

r': distance separating nucleus from the current point of the electron cloud

$\rho(\vec{r}')$: electron concentration at point $\vec{r}'$ of electron cloud

Ω: volume of electron cloud

Ze: electron charge of core (Coulomb)

ε_o: vacuum permittivity

This screening (shielding) effect can be assessed by means of quantum chemistry [VES 99]. The space is split into cubes of equal sizes, the tops of which are the nodes of a three-dimensional net. Using these points, we then calculate both the rigorous potential of the quantum chemistry and the potential corresponding to the first term of V_{elec}. The ratio of these values provides the Z_e' value of the effective nuclear charge. Note that X-ray diffraction is an alternative method to quantum chemistry.

Ultimately, the V_{elec} potential becomes

$$V_{elec,i,j} = \frac{Z_i' Z_j' e^2}{4\pi\varepsilon_o r_{i,j}}$$

2) Dispersion term:

$$V_{VdWi,j} = \varepsilon_{i,j}\left[\left(\frac{r_{ij}^*}{r_{ij}}\right)^{12} - 2\left(\frac{r_{ij}^*}{r_{ij}}\right)^{6}\right] \qquad \text{(Lenard-Jones potential)}$$

Energy ε_{ij} is the Van der Waals potential well depth, that is, the minimum of this potential, and r_{ij}^* is the distance corresponding to this minimum. To be more precise, we can write out [NEM 83]:

$$V_{VdWi,j} = \frac{A_{i,j}}{r_{ij}^{12}} - \frac{B_{i,j}}{r_{ij}^{6}}$$

Coefficients A_{ij} and B_{ij} depend on the nature of the atoms in question.

$$A_{ij} = \varepsilon_{ij}(r_{ij}^*)^{12} \; ; \; r_{ij}^* = \frac{1}{2}(r_{ii}^* + r_{jj}^*)$$

$$B_{ij} = \frac{3}{2}\left(\frac{eh}{2\pi m_e^{1/2}}\right)\frac{\alpha_i\alpha_j}{(\alpha_j/N_j)^{1/2} + (\alpha_i/N_i)^{1/2}} \qquad \varepsilon_{ij} = \frac{B_{ij}}{2(r_{ij}^*)^6}$$

α_i, α_j: polarizability of atoms i and j determined by experimentation

N_i, N_j: number of electrons effectively surrounding atoms i and j

h: Planck's constant

m_e: Electron mass

Some authors express the V_{VdWij} potential by means of an exponential as follows:

$$V_{VdW\,i,j} = A\exp(-Br_{ij}) - \frac{C}{(r_{ij})^6}$$

Gavezzotti [GAV 94] provides the value of these parameters according to the nature of the atom in question. This law was proposed by Buckingham and Corner [BUC 47] with the expression:

$$B_{ij} = \frac{1}{2}(B_{ii} + B_{jj}) \quad \text{and} \quad C_{ij} = (C_{ii}C_{jj})^{1/2}$$

3) Hydrogen bond:

This is the bond of an atom to a hydrogen atom.

According to Lifson *et al.* [LIF 79]:

$$V_H = \frac{A}{r^9} - \frac{C}{r^6}$$

According to Agler *et al.* [AGL 79]:

$$V_H = \frac{A}{r^{12}} - \frac{B}{r^{10}}$$

4) Global potential between atoms i and j:

$$V_{ij} = \frac{Z_i' Z_j' e_i e_j}{4\pi\varepsilon_0 r_{ij}} + (1-p)V_{VdWij} + pV_H$$

If atom j is not a hydrogen atom: $p = 0$

If atom j is a hydrogen atom: $p = 1$

For polar molecules, the electron interaction and hydrogen bonds are predominant relative to the Van der Waals forces.

5) Crystallization energy:

For molecule k within a crystal, let us consider a sphere, the radius of which is 3 or 4 mm and centered on molecule k, where n is the number of molecules within this sphere. The bonding energy of molecule k to the crystal is

$$U_k = \sum_{\ell=1}^{n} U_{k\ell}$$

If the crystal lattice comprises several molecules, we must use the arithmetic mean of the energies corresponding to the molecules of the simple crystal lattice. The lengths of the chemical bonds can be found in the work of Stewart [STE 90].

1.1.2. *Attachment energy and layer energy*

Crystallization energy is the total of the two terms:

$$E_{cris} = E_{cou} + 2E_{fix} \quad [1.1]$$

Indeed, let us "peal" a crystal face by removing a layer of molecules of *minimum* thickness $d_{hk\ell}$, so that the layer obtained allows the crystal to be reproduced by simple orthogonal translation of itself. In order to detach the unit of surface of such a layer, we would need to expend the energy equivalent to the attachment energy E_{fix}, which is calculated by considering the molecules situated in a hemisphere centered on the molecule in question, with a radius of 3–4 mm.

On the other hand, each molecule of the layer is bonded to the other molecules of the same layer. The corresponding energy calculation E_{cou} is done by considering the molecules situated in a radius of 3 or 4 mm.

Within a crystal, each layer is in contact with the rest of the layers in the crystal on both faces, hence the coefficient 2 in equation [1.1].

Equation [1.1] shows that when considering the faces directed in different ways, energies E_{cou} and E_{fix} *vary in opposite ways* due to their total remaining constant and equal to E_{cris}.

Energies E_{fix} are negative and in the order of 100 $MJ.kmol^{-1}$.

1.1.3. *Superficial and molecular energies*

Where n_0 is the number of molecules contained per square meter of layer, the relationship between the energy per surface unit E_s and the energy per molecule E_m is:

$$E_{sfix} = n_0 E_{mfix} \quad \text{and} \quad E_{scou} = n_0 E_{mcou}$$

This relationship applies to E_{cou} and E_{fix}. However, crystallization energy typically concerns the molecule. However, where N_0 is the number of molecules per volume unit, we have:

$$N_0 E_{mcris} = E_{vcris}$$

where E_{vcris} is the crystallization energy per volume unit.

1.1.4. *Surface energy*

The interfacial crystal–solution energy depends on:

– temperature;

– the characteristics of the solution and of the crystal.

According to Onsager [ONS 44], the linear energy of the edge of a face is as follows:

$$\gamma_\ell = 2J' - kT \coth(J/kT) \qquad (J.m^{-1})$$

where J and J' are the interaction energies in two perpendicular directions.

This energy is cancelled for:

$$T = 2J'\left[k \operatorname{Ln} \coth(J/kT)\right]^{-1}$$

When this temperature is reached, the crystal growth becomes rough. This is thermal transition.

The surface energy is:

$$\gamma = (\gamma_\ell / a) \qquad (J.m^{-2})$$

a: dimension of one molecule (m)

For conditions close to ambience, Mersmann proposed the following correlation for surface energy:

$$\gamma = 0.414 kT \left(\frac{\rho_s N_A}{M}\right)^{2/3} \operatorname{Ln}\left(\frac{c_s}{c_L^*}\right)$$

ρ_s: real density of crystal (kg.m^{-3})

M: molar mass (kg.kmol^{-1})

N_A: Avogadro number (6.02.10^{26} molecules.kmol^{-1})

$$c_s = \rho_s / M$$

c_L^* : molar concentration of the saturation liquor

1.1.5. *Periodic chain and nature of faces*

Within a crystal, molecules are bonded between each other by forces that can be represented by a geometric vector sum of the bonds between the atoms of both neighboring molecules. These intermolecular bonds recur periodically in a direction parallel to the bond considered, forming what is known as a bond chain that is clearly periodic.

Typically, faces are categorized according to the number of periodic chains parallel to them and accounting only for chains that contain strong bonds (for instance, Hydrogen bonds). According to Hartmann and Perdok [HAR 55a, HAR 55b, HAR 55c], we can distinguish:

1) K faces that are not parallel to any periodic chain (PC). K faces develop very quickly.

2) F faces containing at least two PCs. These faces develop slowly and, as a consequence, are dominant from a morphological point of view.

3) Faces containing only one PC whose growth is intermediate.

Note that by saying that a K face has no PC is the same as saying that the E_{cou} energy is low, and consequently, the attachment energy E_{fix} is strong.

A mnemonic illustration of this would be to consider the molecules as the tops of tightly packed cubes. Each molecule is bonded to six neighbors on six sides (more or less common) by the eight cubes sharing the same molecule:

1) K faces: Each cube top is the meeting point of three edges joining the top to tops A, B and C. The plan of triangle ABC defines a K-type face. Each surface molecule of this face has only three bonds, unlike the molecules inside the crystal that have six bonds (which is known as a coordination number of 6). This is why the molecules on the K faces are known as *half-sites*.

2) S faces: The surface is in steps. Each cube top has four bonds on six.

3) F faces: The surface is flat and even. Each surface molecule has a coordination number equal to 5.

1.2. Form

1.2.1. *Crystal system and habit*

The habit (from the Latin: *habitus*, external form) of a crystal is determined by the seven crystal systems to which it belongs (see [BRA 66]).

The occurrence frequency of each system in nature is given in Table 1.1.

Monoclinic	29
Orthorhombic	21
Cubic	15
Triclinic	11
Trigonal	10
Hexagonal	7
Quadratic	4
Total	100

Table 1.1. *Bravais' crystalline systems*

Metals typically crystallize in an isotropic system (that is, the cubic system).

As far as filtration or drying is concerned, crystals can occur in one of the following forms:

– tablet;

– prism;

– acicular (fibers, needles, rods);

– pyramid;

– rhombohedron;

– tetrahedron;

– hexahedron (cubes);

– octahedron;

– dodecahedron;

– scale and plate.

Table 1.2 gives the occurrence frequency of these polyhedra according to the system to which the crystal belongs.

The total for each column is equal to 100.

Attrition is often active in crystallizers. Crystal tops and edges are dulled and their form can approach that of river pebbles.

	Monoclinic	Orthorhombic	Cubic	Triclinic	Trigonal	Hexagonal	Quadratic
Tablet	42	31		43	30	37	32
Prism	36	46		29	26	42	29
Acicular	15	19		12	15	11	
Pyramids		4		5		10	24
Rhombohedron				9	29		
Tetrahedron			10				
Hexahedron			31				
Octahedron			38				
Dodecahedron			21				
Scale	7			2			15

Table 1.2. *Practical shapes of crystalline systems*

1.2.2. *Non-sphericity index (form index)*

This is the ratio of the crystal's exterior surface to the surface of the sphere of the same volume.

1) Cube of side L:

Volume $$L^3 = \frac{\pi d^3}{6} \qquad d = \left(\frac{6}{\pi}\right)^{1/3} L$$

Sphere surface $$\pi d^2 = \pi L^2 \left(\frac{6}{\pi}\right)^{2/3}$$

Cube surface $$6L^2$$

$$I_{ns} = \frac{6}{\pi}\left(\frac{\pi}{6}\right)^{2/3} = \left(\frac{6}{\pi}\right)^{1/3} = 1.24$$

2) Parallelepiped of sides A, B and C:

Volume $$ABC = \frac{\pi d^3}{6} \qquad d = \left(\frac{6}{\pi}ABC\right)^{1/3}$$

Sphere surface $$\pi d^2 = \pi\left(\frac{6}{\pi}ABC\right)^{2/3}$$

Parallelepiped surface $$2(AB + BC + AC)$$

$$I_{ns} = \frac{2}{\pi}\frac{(AB + BC + AC)}{\left(\frac{6}{\pi}ABC\right)} \times \left(\frac{6}{\pi}ABC\right)^{1/3}$$

$$I_{ns} = \frac{1}{3}\left(\frac{1}{C} + \frac{1}{B} + \frac{1}{A}\right) \times \left(\frac{6}{\pi}ABC\right)^{1/3}$$

3) Rod of length L and cross-section S (perimeter P):

Volume $\quad SL = \frac{\pi d^3}{6} \qquad d = \left(\frac{6}{\pi}SL\right)^{1/3}$

Sphere surface $\quad \pi d^2 = \pi\left(\frac{6}{\pi}SL\right)^{2/3}$

Parallelepiped surface $\quad LP$

$$I_{ns} = \frac{LP}{\pi} \times \left(\frac{\pi}{6SL}\right)^{2/3}$$

If $P = \pi D$ and $S = \pi D^2/4$

$$I_{ns} = LD \times \left(\frac{2}{3D^2L}\right)^{2/3} = \left(\frac{2}{3}\right)^{2/3} \times \left(\frac{L}{D}\right)^{1/3}$$

4) Plate of thickness E and surface S:

Volume $\quad SE = \frac{\pi d^3}{6} \qquad d = \left(\frac{6}{\pi}SE\right)^{1/3}$

Sphere surface $\quad \pi d^2 = \pi\left(\frac{6SE}{\pi}\right)^{2/3}$

Plate surface $\quad 2S$

$$I_{ns} = \frac{2S}{\pi}\left(\frac{\pi}{6SE}\right)^{2/3} = \left(\frac{2}{\pi}\right)^{1/3}\left(\frac{S^{1/2}}{3E}\right)^{2/3}$$

If $S = \pi D^2/4$

$$I_{ns} = \left(\frac{2}{\pi}\right)^{1/3}\left(\frac{\pi}{4}\right)^{1/3} \times \left(\frac{D}{3E}\right)^{2/3} = \frac{1}{2^{1/3}3^{2/3}} \times \left(\frac{D}{E}\right)^{2/3}$$

NOTE.– Use of I_{ns}:

We always know the mass of the crystals in question and, by knowing their real density ρ_s, we know their real volume. Below, we show that when treating a more or less dispersed population, and when the surface/volume ratio must be retained, the characteristic dimension is the harmonic mean of the different classes of the particle size distribution:

$$\frac{M}{L} = \sum_i \frac{m_i}{L_i} \quad \text{with} \quad M = \sum_i m_i$$

In addition, if we know the form of these crystals, we can deduce their mean dimensions and thereby learn the value of the diameter d and that of volume ω_s of the equivalent sphere:

$$\left(\omega_s = \frac{\pi d^3}{6}\right)$$

The total surface of the equivalent population is:

$$S_T = I_{ns}\left(\frac{M}{\rho_s \omega_s}\right) \times \pi d^2 = I_{ns} \times \frac{M}{\rho_s} \times \frac{6}{d}$$

It is important to know the total surface in crystallization, filtration and drying operations.

The non-sphericity index can reach a value in the order of 10 for an acicular or thin plate. Typically, it falls between 1 and 2.

Coulter counter readings give a direct value for the crystal volume. Equivalent sphere diameter d is immediately deduced and by knowing the crystal form, that is, I_{ns}, we obtain the surface.

1.2.3. *Nyvlt coefficients*

This author avoids using the mediation of the equivalent sphere. Usually, crystals have three main dimensions A, B and C, so that:

$$A \geq B \geq C$$

The intermediate dimension B is the only dimension that can be correlated with granulometric analysis by sifting. Consequently, we define the coefficients directly:

$$\alpha = \Omega / B^3 \qquad \beta = \Sigma / B^2$$

Ω and Σ are the volume and the surface of a crystal taken at random. All of the crystals are supposed to be geometrically similar, which ensures the independence of α and β relative to the crystal size.

If the form of the crystals changes through growth, then the growing rate will be:

$$G = \frac{1}{t}\left[\left(\frac{\Omega}{\alpha}\right)^{1/3} - \left(\frac{\Omega_0}{\alpha_0}\right)^{1/3}\right] = \frac{B - B_0}{t}$$

Note that if R is the rate of orthogonal growth for a face perpendicular to dimension B, the crystal growth rate is:

$$\frac{dL}{dt} = G = 2R$$

1.2.4. *The importance of form: porosity*

The porosity ε of a crystal bed is a direct function of the form.

Non-pressed loose fibers	$0.9 < \varepsilon < 0.99$
Loose plates	$0.80 < \varepsilon < 0.9$
Irregular but equidimensional particles	$0.4 < \varepsilon < 0.7$

Spheres	$\varepsilon \# 0.4$
Tight parallel fibers	$\varepsilon \# 0.2$

A minimum volume is required for explosives. It is advantageous that these particles be spherical.

Silver halide used for photography takes the form of plates due to the very high non-sphericity coefficient for this configuration. These halides offer a maximum surface volume and, for a given quantity of silver, optimal light sensitivity.

1.2.5. *Orthogonal rate and form development*

The growth of a face is characterized by its growth rate to the face in question. This is the orthogonal rate R of the face.

Now let us consider two faces A and B that intersect along edge O, examining the rate R on a plane perpendicular to this edge.

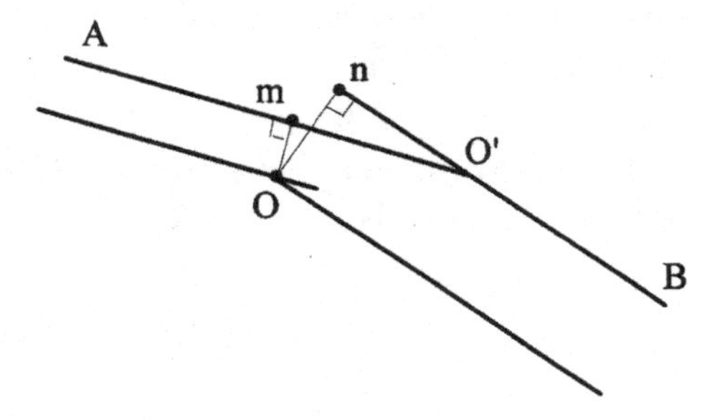

Figure 1.1. *Development of crystal faces*

The orthogonal growth of faces A and B are, respectively, Om and On with, as in the figure:

$$(Om)_A < (On)_B \quad \text{so} \quad R_A < R_B$$

We can see that face A has increased its surface by mO', whereas face B has lost nO'. The edge has moved from O to O'.

Face A becomes *important* and, after sufficient time, face B will disappear, thereby becoming a *virtual* face. Corresponding with the growth progression of the crystal, the number of faces decreases, becoming simpler in form. *Important faces have the lowest orthogonal rates.*

Here, we can introduce the concept of a face's *morphological importance* (MI) as the statistical mean of a face's surface compared with the mean of the crystals' total exterior surface. It is consistent to establish MI for a crystal population close in size, since their form increasingly varies as size increases.

1.2.6. *Form modifier additives*

Additives slow the growth of certain faces. They strongly attach to the face in question, and effective additives must have significant bonding energy with the crystal.

In order to attach themselves to the crystal, additives must be "tailor made", which essentially means being a crystal molecule that has been modified so that the modified part emerges from the surface so as to prevent other molecules from attaching, thereby reducing the growth of the face in question.

Additives act differently according to whether they are *cutting* or *blocking*.

Cutting additives alter the sequence of bonds or, more precisely, cut the bond between the face on which they are attached and the next layer, as the additive is unable to bond with molecules of this layer. Let us examine the two modes of action of cutting additives. Some use an oxygen atom that takes the place of a hydrogen atom. The hydrogen atom bonds with the oxygen atom of molecules attaching to it. With this oxygen atom, bonding is no longer possible, being replaced by repulsion between the two oxygen atoms. The H – O bond is *cut*.

Similarly, if the normal molecule has an oxygen atom susceptible to bond to the hydrogen atom of a molecule fixing on it, we replace the oxygen atom with the hydrogen atom (brought by CH or NH), and the O – H bond is cut.

The emerging part of a blocking additive has a greater volume than that of normal molecules, which prevents the next layer from attaching. This emerged part can be a large chlorine atom, a methyl radical, a benzene ring, a naphthalene radical or even anthracene.

Paraffin typically crystallizes as plates, with the additive attaching to the edges of these plates that are then forced to increase in thickness. We then obtain crystals that are strongly equidimensional in character, that is, their three dimensions are of the same order. The specific surface of these crystals is lower, which means that they can be more readily obtained by filtration.

The insertion energy of *a single* additive molecule is the total:

– of the additive's bonding energy with the crystal surfaces *below*;

– of the bonding energy with the molecules of the layer to which the additive belongs.

$$E_{ins} = E_{sous} + E_{cou}$$

Here, all energy is counted positively (despite being a potential well in this case).

Once the insertion energy E_{ins}^{A} for an additive has been calculated, together with that for a normal molecule E_{crist}^{N}, we can calculate:

$$\Delta E = E_{ins}^{A} - E_{crist}^{N}$$

The additive receptor face(s) are those for which:

$$\Delta E \geq 0$$

These calculations are performed by assigning a potential to each atom (see section 1.1.1). ΔE increases with the molecular mass of the additive.

Both the size of the metastable zone and the crystallization induction time increase with the additive's insertion energy.

1.3. Conclusions: crystal characteristics

1.3.1. *Form and specific surface*

Form is announced by a more or less equidimensional character of the crystal, that is, when the three main dimensions of an equidimensional crystal are not dissimilar. In the order of increasing non-sphericity, we can distinguish:

– equidimensional crystals: pyramids, rhombohedra, tetrahedra, hexahedra (cubes), octahedra (cube with truncated corners), dodecahedra;

– one dimension is slightly lower than the others (tablets);

– one dimension is significantly lower than the others (plates, scales);

– one dimension is slightly greater than the others (prisms);

– one dimension is significantly greater than the others: acicular (rods, needles, fibers).

Specific surface σ increases with non-sphericity.

A high specific surface is:

– detrimental for permeability (filtration);

– favorable for drying;

– favorable for chemical reactivity.

Specific surface is acquired directly by filtration.

The intermediate dimension of the crystal is acquired by screening.

The volume is provided by a Coulter counter.

Finally, by microscopic examination of fifteen crystals, we can fix a mean for the ratios:

$$L_1/L_2 \text{ and } L_3/L_2 \quad \text{with} \quad L_1 < L_2 < L_3$$

Knowledge of these two ratios allows us to evaluate the α and β Nyvlt coefficients in order to calculate both volume and surface:

$$V = \alpha L_2^3 \qquad\qquad S = \beta L_2^2$$

Specific surface σ can be estimated easily if the particles can be assimilated to solids of a simple geometrical form. For example:

– sphere of diameter d

$$\sigma = \pi d^2 / \left(\frac{\pi d^3}{6} \right) = 6/d$$

– fiber of diameter d and length L

$$\sigma = L\pi d / \left(\frac{L\pi d^2}{4} \right) = 4/d$$

– plate of thickness e and surface S

$$\sigma = 2S / Se = 2/e$$

1.3.2. *Chemical composition: hygroscopicity*

On the surface, hydrogen atoms are provided by hydroxyl, amine and imine radicals. However, carbonyl groups are hydrogen acceptors.

Consequently, we observe that the water molecules are attracted:

– in the first instance, by their oxygen atom that bonds to the surface hydrogen atoms;

– in the second instance, by their hydrogen atoms that bond to the carbonyl groups.

Therefore, soda crystals are highly hygroscopic due to their hydroxyl radicals.

More generally, a crystalline surface can be polarized positively or negatively and thereby attract polarized molecules, one of which could be water for instance.

1.3.3. *Surface striations*

A series of steps can occur successively on a given face F.

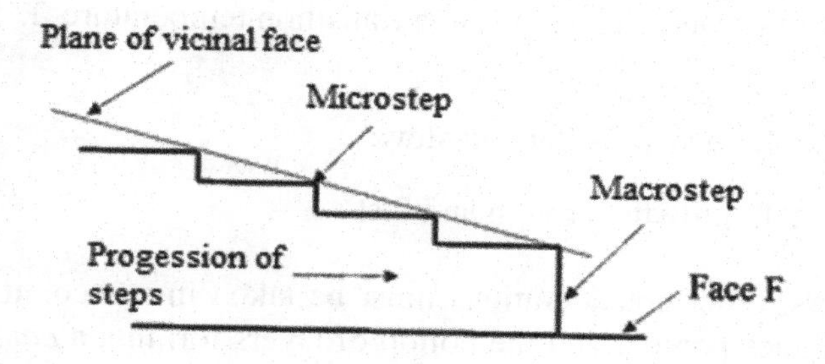

Figure 1.2. *Macrostep and vicinal face*

The rate of advance for successive simple steps is roughly the same globally due to their height being comparable to the size of a single molecule. They can reach the height of a macrostep, which is considerably greater in height, and consequently, slower in advance rate. There are also additives that slow and even stop macrostep progression [VAN 86]. The state of the surface resulting from this grouping of steps presents striations that are clearly visible, corresponding to these macrosteps. However, if the microsteps are not visible, the surface will still present a slope that is very low in gradient relative to the ideal F face. Such a surface is known as a *vicinal* face, since its orientation is close to that of face F (from the Latin *vicinus*: neighbor).

1.3.4. *Surface roughness*

A rough surface can be of interest for chemical reactivity and dissolution speed. Similarly, crystals attach to one another more readily without slipping, thereby ensuring greater cohesion for tablets.

However, broadly speaking, the rate of growth for a rough face is similar to that of a K-type face, that is, very high. As a consequence, the morphological significance of rough faces is low. These faces can become virtual.

Bennema and his team [GRI 98] explained why certain F faces of rough character can have fast growth. According to Onsager's bidimensional theory [ONS 44], the *linear step energy* $(J.m^{-1})$ *decreases linearly with the temperature* and is cancelled by a rough transition temperature T_c, known as the Ising temperature.

– where $T < T_c$ growth is flat and slow;

– where $T \geq T_c$ growth is rough and fast.

Nonetheless, other considerations must be taken into account. Bennema and his colleagues popularized the notion of layers forming a *connected net.* This net is obtained by the combination of the two non-parallel bonding chains. However, unlike a classic layer, a connected net is not necessarily stoichiometric, which means that polar net can exist.

A *step* corresponds to the development of a connected net comprising the breaking of bonds on its front, and the apparition of new bonds behind this front. When the difference between the two is zero, the linear energy γ is also zero, and the face can develop very quickly irrespective of temperature. Among the F faces able to develop quickly, the authors distinguish:

– rough faces, where γ is zero irrespective of the temperature;

– disordered faces, where two connected nets have different transition temperatures;

– flat but reconstructed faces, where the influence of the mother liquor can lead to rough growth.

Meekes *et al.* [MEE 98] explained the notion of roughness through symmetry $(\gamma = 0)$, while Grimbergen *et al.* [GRI 99] studied multiple connected nets. This is as far as we will discuss this subject.

1.3.5. *Attrition on drying: vitrification on grinding*

Subject to dryer design, crystals can be eroded during drying operations and become rounded as a consequence. Accordingly, a tablet would take a cushion or rounded almond form with a rough surface from a microscopic perspective.

Sugar crystals are susceptible to vitrification when finely ground, but their surface will be smooth.

1.3.6. *Surface concavity*

Crystals only present concavities if there are dendrites or macles (interpenetration of two crystals) present or at the beginning of dissolution.

1.3.7. *Color*

The color of a crystal is due to:

– the presence of specific radicals in the molecules;

– the disposition and nature of metallic ions that it contains (chrome yellow, cobalt blue, iron oxides beige and brown, titanium white and, though not a metal, carbon black).

Recall that the pigments (such as those mentioned above), like a great majority of minerals, are insoluble in water, unlike colorants that are organic molecules and soluble in water (chlorophyll, hemoglobin, to cite only two colorants). Certain colorants are used, together with pigments, in plastic materials, but they are expensive.

A measurement of sugar coloration is obtained by the absorption of light with a wavelength of 420 nm.

1.3.8. *Flow behavior*

The propensity of a divided solid to flow freely depends on:

– particle form;

– solid porosity.

The flow becomes more difficult as the form of the crystals becomes less spherical. The rate of non-sphericity is 1 for a sphere, reaching 5 for plates and 10, 50 or more for fibers or acicular forms. Angular particles flow poorly.

A divided solid with a loose structure, that is, with high porosity, flows more readily than a compacted solid (for example, by vibrations or repeated impacts) that may not even flow at all.

1.3.9. *Purity*

Impurities can be present in two ways, both of which are often unpredictable:

– inclusions,

– coating.

Impurity inclusions or even mother liquor inclusions are of greater concern when the crystal growth rate is higher.

Coating is due to the presence of mother liquor on the crystal surface, which has not been sufficiently drained by filtration. In addition, we should recall that *tailor-made* additives, made to modify the form of crystals, are located on the crystal surface. Any impurities present in the industrial solutions or in the atmosphere often have an unpredictable effect.

When, as is often the case, the impurity is more soluble in the mother liquor than the material that we wish to crystallize, this impurity becomes concentrated in the liquid, allowing for its elimination by a purge, the flow rate of which must be calculated judiciously.

1.3.10. *Elasticity constants*

In deformable solid mechanics, it is demonstrated that the deformation tensor and stress tensor are symmetrical. These tensors are of rank 2 and order 3, and can be written as:

$$\varepsilon_{k\ell} = \varepsilon_{\ell k} \quad \text{and} \quad \sigma_{ij} = \sigma_{ji}$$

Accordingly, we can number the indexes in the following manner:

ij	11	22	33	23	31	12
q	1	2	3	4	5	6

However, the theory of linear elasticity is written strictly as:

$$\sigma_{ij} = E_{ijk\ell}\varepsilon_{k\ell}$$

$E_{ijk\ell}$ is a rank 4 tensor.

Using the notations above, we can simply write out:

$$\sigma_q = E_{qr}\varepsilon_r \quad [1.2]$$

We will show that matrix E_{qr} is symmetrical.

For two stress-deformation couples, the Helmoltz energy of the solid varies in the following manner:

$$dU = \sigma_i d\varepsilon_i + \sigma_j d\varepsilon_j \quad [1.3]$$

Let us introduce potential Φ (that is a state function):

$$\Phi = U - \sigma_i\varepsilon_i - \sigma_j\varepsilon_j \quad [1.4]$$

By combining [1.3] and [1.4], we obtain:

$$d\Phi = -\varepsilon_i d\sigma_i - \varepsilon_j d\sigma_j$$

However, Φ is a state function. Consequently:

$$\sigma_i = -\frac{\partial\Phi}{\partial\varepsilon_i} \quad \text{and} \quad \sigma_j = -\frac{\partial\Phi}{\partial\varepsilon_j}$$

Hence:

$$-\frac{\partial^2\Phi}{\partial\varepsilon_i\partial\varepsilon_j} = \frac{\partial\sigma_i}{\partial\varepsilon_j} = \frac{\partial\sigma_j}{\partial\varepsilon_i}$$

In other words, the elasticity constant matrix E_{ij} is a symmetrical matrix of rank 2 and order 6. Therefore, its maximum number of elements is 21.

$$E_{ij} = E_{ji}$$

The E_{ij} elements of this matrix have properties that depend on the symmetrical properties of the crystal in question [NOV 95].

Therefore, for the triclinic crystal system with the lowest number of symmetries, the 21 elements are all different. On the other hand, for the cubic system, only three elements are required to characterize a crystal's elastic properties. These elements are E_{11}, E_{44} and E_{12}.

$$\begin{matrix} E_{11} & E_{12} & E_{12} & 0 & 0 & 0 \\ & E_{11} & E_{12} & 0 & 0 & 0 \\ & & E_{11} & 0 & 0 & 0 \\ & & & E_{44} & 0 & 0 \\ & & & & E_{44} & 0 \end{matrix} \qquad \begin{matrix} E_{11} = E_{22} = E_{33} \\ E_{12} = E_{21} = E_{13} = E_{31} \\ E_{44} = E_{55} = E_{66} \end{matrix}$$

A complete table of the various crystal systems may be found in the work of Novick [NOV 95].

These *Young models* are measured in GPa (1 gigaPa = 10^9 Pa) and, for a perfect crystal, their value is

$$-1\,\text{GPa} < E_{ij} < 200\,\text{GPa} \qquad \text{[SAN 87]}$$

Sanquer *et al.* [SAN 87] established that there is identity between the measured values of sound wave velocity and those calculated based on crystal potential.

This calculation is based on the fact that a displacement has two consequences:

– the apparition of deformation tensor;

– the apparition of a modification in crystal potential with the apparition of stress tensor as a corollary.

For organic compounds, E_{ij} does not exceed 30 GPa but can reach 200 GPa for metals such as iron or for minerals.

1.3.11. *Hardness, deformation and fracture*

In the broadest sense, crystals deform by slipping, with hard crystals being more resistant to deformation. Typically, deformation occurs where the potential hollow to be crossed is shallowest.

A good parameter to estimate the resistance of a crystal plane to slipping is the following:

$$\frac{a}{d_{hk\ell}} = \frac{\text{intermolecular distance in the plane of slip in question}}{\text{interplanar spacing of this plane}}$$

A low value for this ratio, that is, a high molecular density in the plane (low a) and a high interplanar distance $d_{hk\ell}$, are the characteristics of soft materials. Graphite, talc and soap flakes deform readily, in a similar fashion to the MoS_2 used as a lubricant in oils.

However, in most crystals, such a mechanism would require significant stress, which in reality is not the case. This contradiction is removed by observing that slips require the presence of local defects or corner or spiral dislocations, in which potential hollow crossing occurs more readily than it would in a perfect net [QUÉ 88].

The depth of the potential hollow to be crossed increases with the strength of the intermolecular bonds of the crystal. Hard crystals are crystals with strong bonds.

Therefore, the ionic bonds in ores and the covalent bonds in diamond and silicon are clearly stronger than the Van der Waals bonds present in crystals used by the pharmaceutics industry.

Typically, we observe two sorts of fracture:

– Ductile fracture, which follows an irreversible and often significant plastic displacement. The stress corresponding to traction is moderate in its value.

– Fragile fracture, where the two surfaces of the broken solid may be restored to contact with a near perfect alignment to the atomic level. The stress corresponding to fragile fracture is high in its value.

In reality, fragile fracture occurs earlier than perfect crystal theory would expect. Indeed, this fracture occurs due to existing microcracks and, where the crystal has none, they can be created by high stress, even short in duration (as in the case of glass shattering with an impact).

In practice, hardness can be measured in various ways:

– perforation pressure using an indenter of low cross-section. This is the Vickers hardness test measured in MPa [YOR 83] as follows:

Aspirin	85
KCℓ	174
NaCℓ	208
Sugar	624

– scratching a harder body and/or scratching a softer body with the body in question. In this way, Mohs developed a hardness scale going from 1 to 10 as follows:

Wax	0.02
Graphite	0.5–1
Talc	1
Organic Crystals	2
Glass	4.5–6.5
Alumina	9.25
Diamond	10

Both perforation and scratching correspond to a superficial fracture phenomenon that is:

– ductile for soft bodies;

– fragile for hard bodies.

There is a transition temperature T^* between fragile fracture and, for $T \geq T^*$, ductile fracture. For example (in Kelvin), approximately:

Mild steel	160
NaCℓ	650
Silicon	750

In short, we can describe hardness as an increasing function of elasticity model E.

Hard particles	E from 10^8 to 10^9 Pa
Soft particles	E in the order of 10^5 Pa

1.3.12. *Fragility and ductility of crystals*

From the data provided by Mersmann [MER 01], we can deduce:

1) Fragile crystals:

Crystals with covalent bonds are typically hard and fragile, as these are directional bonds that hinder the reticular plane from slipping.

$E / H_v < 50$ and $H_v > 300$

2) Ductile crystals:

Ionic bonds are not directional, which encourages slipping of reticular planes.

$E / H_v > 200$ $\qquad H_v < 200$

On collision with an agitator or pump rotor, the stress application speed is significantly greater than it would be during a Vickers hardness test, which takes place at near static speeds. Consequently, plastic deformation is hindered.

1.3.13. *Agglomeration in suspension*

Agglomeration is a result of collisions between particles. It is most significant for crystals smaller than 2 or 3 μm, becoming negligible for crystals of 30 μm.

The degree of agglomeration Z is the number of crystals present in an agglomerate. A degree of 80 has been observed. As a general rule, agglomeration only applies to crystals derived from primary nucleation. Indeed, secondary nucleation results in agitation which destroys agglomerates.

The force of interaction between the two particles is:

$$E = \pi r\left(-\frac{A}{12\pi D} + \frac{64 N_{oi} kT\Gamma_o^2 \exp(-\kappa D)}{\kappa^2}\right) = E_{attrV.d.W} - E_{repulsion}$$

D: distance between the surfaces of two particles

A: Hamaker constant (10^{-18} J)

r: particle radius (m)

Debye–Hückel distance is $1/\kappa$ with:

$$\kappa = \sqrt{\frac{\sum_i (z_i e)^2 N_{oi}}{\varepsilon\varepsilon_o kT}}$$

N_{oi}: concentration of type i ions ($kmol.m^{-3}$)

Γ_o contains the surface potential Ψo

$$\Gamma_o = \frac{\exp(ze\Psi o / 2kT) - 1}{\exp(ze\Psi o / 2kT) + 1}$$

Attractive forces exist between hydrophobic surfaces. One known example of this is the micelle formation of surfactants. On the other hand, solvated particles repel each other.

The aggregation of silica is a classic example.

$$SiO_3Na_2 + 2H^+ \rightleftharpoons SiO_2 + H_2O + 2Na^+$$

In the basic medium, particles are negatively charged and do not agglomerate. However, if the concentration of electrolyte increases, surface charges are compensated and aggregation begins. However, aggregation is fast at $pH < 7$ due to the monosilicic acid neutralizing the OH^- ions, which were absorbed on the crystals. The concentration acts by means of the exponential of $(-\kappa D)$.

According to von Smoluchowski, we can choose:

$$\frac{dN}{dt} = -\beta N^2$$

N: particle concentration

Two mechanisms apply: perikinetic and orthokinetic

1) Perikinetic:

$$\beta = 4\pi DL$$

D: diffusivity ($m^2.s^{-1}$)

L : particle size (m)

2) Orthokinetic:

$$\beta = \frac{2}{3} A\dot{\gamma}L^3$$

Note that the solid volume fraction φ is:

$$\varphi = \alpha L^3 N$$

Hence, replacing L^3N by its value, the aggregation equation becomes:

$$\frac{dn}{dt} = -\frac{2}{3\alpha} A\dot{\gamma}\varphi n$$

The decrease in n is exponential.

$\dot{\gamma}$: shear rate (s^{-1} $(\dot{\gamma} \sim N)$)

n : number of particles per cubic meter

N: here, motor rotation frequency ($turn.s^{-1}$)

β is *inversely proportional to the suspension's solid load* ($kg.m^{-3}$) due to agglomerate fracture.

Oversaturation consolidates the agglomerates.

Perikinetic aggregation becomes orthokinetic aggregation for crystal sizes from 10 to 30 μm.

Ultimately, agglomeration is favored by:

– high levels of diffusion;

– low viscosity. Indeed, diffusion is provided by the Einstein formula:

$$D = \frac{kT}{3\pi\mu d_p} \qquad (m^2.s^{-1})$$

μ: viscosity (Pa.s)

d_p: particle diameter (m)

k: Boltzmann constant ($1.38.10^{-23}$ $J.K^{-1}$),

so, high temperature;

– high concentration of particles (number);

– for particle sizes less than 2 or 3 μm, perikinetic agglomeration can be intense.

Agglomerate's resistance to fracture is characterized by the ratio [ORO 49]:

$$\frac{\Gamma}{K} \sim 1.7E\left(\frac{2}{nC_c N_A}\right)^{1/3} \qquad 5\ J.m^{-2} < \frac{\Gamma}{K} < 15\,J.m^{-2}$$

C_c: kilomoles per cubic meter of crystal

N_A: Avogadro number ($6.023.10^{26}$ molecules per kilomole)

n: number of molecules in the agglomerate.

Γ/K is far greater than the superficial energy, which is less than 0.1 $J.m^{-2}$.

The Young modulus for crystals is in the order of:

$$E \# 2.10^{10}\,Pa$$

The fracture resistance of a polycrystal of 1 mm is:

$$\sigma \# 0.005E \# 10^{8}\,Pa$$

The fracture resistance of polycrystals is inversely proportional to their size.

1.4. Attrition

1.4.1. *Crystal hardness*

Vickers hardness corresponds to the force F required to pierce the material. The indenter has the force of a pyramid with an angle of 136° at the summit:

$$H_v = 1.854\frac{F}{d_1 d_2}$$

d_1 and d_2 are the two diagonal angles of the indent left by the indenter.

F is measured in Newtons and H_v in Pascals.

1.4.2. *Fracture resistance*

A simple ratio was proposed by Orovan [ORO 49]:

$$\left(\frac{\Gamma}{K}\right) = 1.7E\left(\frac{1}{nc_c N_A}\right)^{1/3}$$

n : number of atoms in the molecule

c_c : molar concentration in the crystal ($kmol.m^{-3}$)

N_A : Avogadro number

E : elasticity modulus (Pa).

Gahn [GAH 97] proposed a measurement based on the work of plastic deformation.

$$W_{pl} = 0.06\left(\frac{F^{*3}}{H_v}\right)^{1/2}$$

F^* : critical force required for fissure formation.

μ is the shear modulus. Expressed in GPa, its value is between 1 and 20.

$$\mu = \frac{E}{2(1+\nu)}$$

ν is the Poisson modulus, typically close to 0.3.

1.4.3. *Attrition in a stirred vessel [MER 01]*

Collisions with the impeller give rise to fragments whose minimum size is:

$$L_{min} = \frac{32}{3}\frac{\mu}{H_v^2}\left(\frac{\Gamma}{K}\right)$$

The mass distribution of fragment dimensions is, disregarding $1/L_{max}$ in favor of $1/L_{min}$:

$$q_0(L) = 2.25 L_{min}^{2.25} L^{-3.25}$$

with:

$$\int_{L\,min}^{L\,max} q_0(L)\,dL = 1$$

Attrition reduces the speed of crystal growth that we can write out as:

$$G = G_0\left[1-\eta\left(\frac{L}{L_{max}}\right)^2\right]$$

We propose:

$\eta = 0.15$ for very hard crystals (SiO_2)

$\eta = 0.5$ for less hard crystals (NaCl, KCl, KF, LiCl, LiBr).

Indeed, attrition provokes wear equivalent to negative growth and proportional to the square of the crystal's dimensions.

Note that Mersmann [MER 01, p. 196] proposed an expression for: L_{max}

$$L_{max} = \frac{1}{2}\left(\frac{H^{2/3}K}{\mu\Gamma}\right)^{1/3} W_{pl}^{4/9}$$

W_{pl} is the energy of plastic deformation

$$W_{pl} = \frac{\pi}{8} H_V a_\pi^3$$

a_π : radius of the area of plastic deformation (m)

1.5. PSD measurement of a crystalline population

1.5.1. *Sifting*

A series of superimposed sieves with decreasing downward apertures is stirred electromechanically in order to encourage the decreasing flow pass through each sieve. The rejected material remains on the sieve mesh.

d: the sieve opening is standardized.

$$5.5\ \mu m < d < 1250\ \mu m$$

The dimension that determines whether or not a particle crosses the sieve mesh is the intermediate dimension.

1.5.2. *The Coulter particle counter*

A suspension of particles in the electrolyte is sucked in through a cylindrical orifice. Then, for each particle, we detect the resistance variation between two electrodes placed on either side of the orifice. The calculation is based on two hypotheses:

– 1st hypothesis: the particle has a length of x and a cross-section of kx^2;

– 2nd hypothesis: current transmission in the orifice is not altered by the particle's presence. $R_0 = \rho L/S$ is the resistance of the liquid channel without the particle.

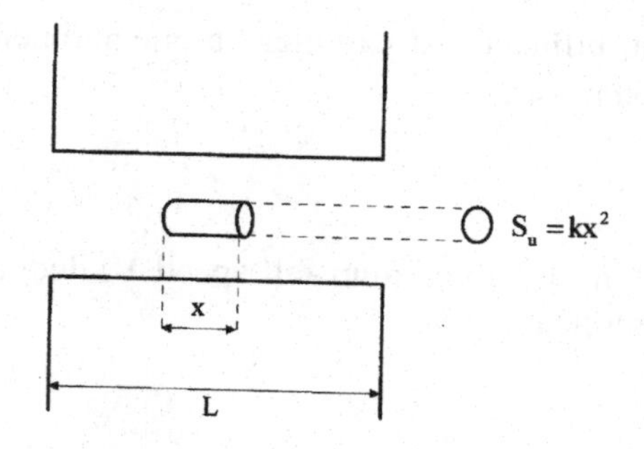

Figure 1.3. *Principle of the Coulter counter*

ρ: resistivity of the liquid (much lower than that of the crystal).

With the solid's presence, resistance becomes:

$$R_1 = \rho\frac{L-x}{S} + \rho\frac{x}{S-kx^2}$$

$$R_1 - R_0 = \Delta R = \rho\left[\frac{x}{S-kx^2} - \frac{x}{S}\right] = \rho\left[\frac{kx^3}{S(S-kx^2)}\right]$$

If $kx^2 << S$ so $\Delta R \sim k'x^3 = KV$

A pulse of resistance, and thereby voltage, corresponding to the passage of a particle, is proportional to the particle's volume.

We see that if we calibrate the device by comparison with measurements performed by microscopy or sifting, we can immediately obtain the volume distribution of a particle population.

The minimum size detectable is in the order of 10 μm.

1.5.3. *Sedimentation*

In order to apply Stokes' law:

$$V = \frac{\Delta\rho g d^2}{18\mu}$$

There must be no influence of particles on one another. Accordingly, the solid volume fraction must be:

$$\varphi < 2.10^{-3}$$

We can then, taking the maximum fall speed V, deduce the particle size d_p. This method is applicable for:

$$2\,\mu m < d_p < 100\,\mu m$$

1.5.4. *Image analysis*

Under a microscope, we sweep over the sample in the light of a laser. The image is then projected onto a television screen on which light areas represent the crystals. Granulometry is provided by means of a software. The laser can be used up to 32% of crystals [WIT 92].

1.6. Characteristics of a crystal population

1.6.1. *Analytic expressions of particle size distribution*

We retain the distribution of both Gauss and Rosin–Rammler.

1) The frequency density f(x) of Gaussian distribution is given by:

$$f(x) = \frac{1}{\sigma\sqrt{2\pi}} \exp - \left[\frac{(x-\bar{x})}{2\sigma^2} \right] \quad \text{with} \quad \int_{-\infty}^{+\infty} f(x)dx = 1$$

σ is the distribution's *standard deviation.*

The *descending fluid* relative to size x includes particles of size smaller than x.

$$P(x) = \int_0^x f(x)dx = \text{erf}(x) \qquad \text{(x's “error function”)}$$

The inverse function of $P(x)$ is $(x) = \mathrm{erf}^{-1}(P) = x_p$

$\overline{x}$ is the mean whereby $P(x) = 0.5$ that is $(\overline{x}) = x_{0.5}$

On the other hand, Gaussian distribution is such that:

$$P(\overline{x}+\sigma) = 0.8413\#0.84 \quad \text{and} \quad P(\overline{x}-\sigma) = 0.157\#0.16$$

and the *variation coefficient* of the distribution is, by definition,

$$CV = \frac{x_{0.84} - x_{0.16}}{2x_{0.5}} \qquad [1.5]$$

For Gaussian distribution:

$$CV = \sigma/\overline{x}$$

2) According to Rosin–Rammler, the rejected material is *given directly* by:

$$R(x) = \exp\left[-\left(\frac{x}{x'}\right)^n\right] \quad \text{with} \quad P(x) = 1 - R(x)$$

Hence:

$$x = x'[\mathrm{Ln}1/R]^{1/n}$$

n is the uniformity parameter;

x' is the characteristic size.

We can directly deduce:

$$\overline{x} = x'[\mathrm{Ln}2]^{1/n}$$

Moreover, according to [1.5]:

$$CV = \frac{\left[\text{Ln}(1/0.16)\right]^{1/n} - \left[\text{Ln}(1/0.84)\right]^{1/n}}{2\left[\text{Ln}2\right]^{1/n}}$$

Indeed, by convention, we have applied the definition of CV given by equation [1.5] to all the distributions.

1.6.2. *Mean size and solubility*

For highly soluble crystals, the mean size is in the order of several hundred micrometers, and with the increasing oversaturation, this size becomes independent of the oversaturation σ due to the preferential attrition of large crystals.

Of less soluble crystals, typically $\sigma \gg 1$, with primary nucleation dominating. Then, aggregation appears, providing the "visible seeds" and particles of several dozen micrometers (however, only where repulsive forces do not intervene). Note that oversaturation must not be too high, since intense nucleation causes it to drop significantly and the agglomerates would only be weakly bonded due to the absence of any strong bonds.

1.6.3. *Coefficient of variation and attrition*

Attrition influences the coefficient of variation CV, that is, the spread of the grain size distribution.

Resistant particles, though principally the largest of them, are abraded by agitation. Spread is reduced, becoming more limited at a specific agitation energy in the order of 1 $W.kg^{-1}$ (per kilogram of suspension).

On the other hand, fragile particles are all abraded and eroded with production of particles between 1 and 150 μm, which significantly extends the range of sizes. However, if agitation increases too much, large particles will also be broken and the size distribution will again be narrowed.

2

Crystal Formation and Growth

2.1. Crystal formation

2.1.1. *Primary nucleation*

All solutions contain agglomerates of solute that are increasingly numerous and significant on approaching the saturation temperature. These are known as crystal embryos.

As a result of a $\Delta\mu$ supersaturation, the Gibbs energy variation of the embryonic solution (for crystal embryo formation) reaches its maximum. The crystal embryos with the corresponding size are the crystal seeds or nuclei. We should note that it is only by means of statistical fluctuations that the Gibbs energy can increase to its maximum. Next, the Gibbs energy of the seed-solution system naturally decreases as the nucleus increases in its size to form a crystal.

The free enthalpy variation corresponding to the crystal embryo formation of diameter d is the difference between the surface energy created and the over-potential $\Delta\mu$ lost in the solution.

$$\Delta G = \pi d^2 \gamma - \frac{\pi}{6}\frac{d^3}{\Omega}\Delta\mu$$

Ω: molar volume ($m^3.kmol^{-1}$)

γ: superficial energy of the crystal embryo ($J.m^{-2}$).

We refer to the publication of Nielsen et al. [NIE 71] for numerous values of the surface energy γ.

$\Delta\mu$: excess of chemical potential relative to saturation ($J.kmol^{-1}$).

By definition:

$$\Delta\mu = kT(Lna - Lna^{*}) = kTLn(a/a^{*})$$

a : solute activity in solution

a^{*} : solute activity at equilibrium

In crystallization operations:

$$a/a^{*} > 1 \text{ and we assume } a/a^{*} = 1+\sigma \text{ that is, } \sigma = \frac{a-a^{*}}{a^{*}}$$

σ: relative supersaturation

Finally:

$$\Delta\mu = kTLn(1+\sigma) \# kT\sigma \quad (\text{if } \sigma << 1)$$

The maximum G is reached when:

$$\frac{d(\Delta G)}{d(d)} = 2\pi\gamma d - \frac{\pi}{2} d^2 \frac{\Delta\mu}{\Omega} = 0$$

hence:

$$r^{*} = \frac{d^{*}}{2} = \frac{2\gamma\Omega}{\Delta\mu} = \frac{2\gamma\Omega}{kT\sigma}$$

$$\Delta G^{*} = \frac{16\pi\gamma^3\Omega^2}{3(\Delta\mu v)^2}$$

Coefficient v is added. This is the number of ions corresponding to the possible electrolyte dissociation.

In his work, Mutaftschief [MUT 01] gives the following expression for primary nucleation:

$$J_0 = \frac{\alpha^* n_1 q_{re}}{i^*}\left(\frac{\Delta G^*}{3\pi kT}\right)^{1/2} \exp\left(-\frac{\Delta G^*}{kT}\right)$$

q_{re} is the partition function for the replacement of movement characteristics by their equivalents in the agglomerate. The corresponding parameters express transfers by a volume, vibrations by a surface and rotations by a length. In other words:

$$Q_{re} = \frac{1}{\sigma}\frac{(V_{al})agg}{(V_{al})cristal} \qquad \left(\begin{array}{c}\sigma \text{ is in the order of 6 maximum} \\ \text{this is the number of symmetries}\end{array}\right)$$

We assume $q_{re} = Q_{re} / V_{agg}$

If, somewhat audaciously, we accept that $V_{a\ell}$ and ℓ are similar for the agglomerate and the crystal, and if we also accept that V is in the order of molecular volume, we obtain:

$$q_{re} \# 1 / \Omega\sigma$$

The product $\alpha^* n_1$ is the number of molecules that come into collision with the agglomerate per unit of time. Here again, and still in a somewhat risky manner, if we apply the Ranz formula for the exchange of material between the ambient environment (the solution) and a small molecule (the agglomerate), we obtain:

$$\alpha^* n_1 = 4\pi r^{*2}\beta c \quad \text{with} \quad \beta = \frac{D}{\delta} \text{ and } \delta = \frac{d^*}{2} = r^*$$

D: diffusivity ($m^2.s^{-1}$)

c: concentration (molecules.m^{-3})

Hence:

$$\alpha^* n_1 = 4\pi r^* Dc$$

We have written c rather than $(c - c^*)$ as here this is the molecular flow arriving in the agglomerate.

The number of molecules i present in the critical agglomerate is:

$$i^* = \frac{4}{3}\frac{\pi r^{*3}}{\Omega}$$

The first factor of the Mutaftschief formula is:

$$\frac{\alpha^* n_1 q_{re}}{i^*} = \frac{4\pi r^* Dc}{\sigma\Omega} \times \frac{3\Omega}{4\pi r^{*3}} = \frac{3Dc}{\sigma r^{*2}}$$

And finally:

$$J_0 = \frac{3Dc}{\sigma r^{*2}}\left(\frac{\Delta G^*}{3\pi kT}\right)^{1/2} \exp\left(-\frac{\Delta G^*}{kT}\right) \quad \text{with} \quad r^* = \frac{2\gamma\Omega}{\Delta\mu} \quad \text{and}$$

$$\Delta G^* = \frac{16\pi\gamma^3\Omega^2}{3(kT\sigma)^2}$$

which is equivalent to:

$$J_0 = \frac{D_c N_A \sigma}{6\Omega}\left(\frac{kT}{\gamma}\right)^{1/2} \exp\left(-\frac{\Delta G^*}{kT}\right)$$

Primary nucleation can only occur for a relative supersaturation σ greater than or equal to 0.5.

Example.–

$D = 10^{-9}\, m^2.s^{-1}$ $\quad \Omega = 1.25.10^{-28}\, m^3.molecule^{-1}$ $\quad \sigma = 1$

$c = 2\ kmol.m^{-3}$ $\quad a = 5.10^{-10}\, m$ $\quad T = 300\, K$

$k = 1.38.10^{-23}\, J$ $\quad N_A = 6.023.10^{26}\, molecule.kmol^{-1}$ $\quad \gamma = 0.024\ J.m^{-2}$

$$\frac{\Delta G^*}{kT} = \frac{16\times\pi\times(0.024)^3\times(1.25.10^{-28})^2}{3\times(1.38.10^{-23}\times300)^3} = 51$$

$$J_0 = \frac{10^{-9}\times2\times6.023.10^{26}}{6\times1.25.10^{-28}}\times\left(\frac{1.38.10^{-23}\times300}{0.024}\right)^{1/2}\exp(-51)$$

$$J_0 = 4.733.10^{13}\,\text{nuclei.m}^{-3}.\text{s}^{-1}$$

2.1.2. *Crystallizer crust formation*

Using a mother liquor to obtain crystals often requires this liquid to be cooled through a metal. Occasionally, solubility decreases with temperature and it must be subsequently heated. A phenomenon known as scaling them occurs on the boiler vapor tubes.

When the difference in temperature between the liquor and the metal surface exceeds a certain threshold (between 2 and 20°C), crystals begin to form on this surface. This is known as crusting, which is highly problematic for the thermal transfer as the thermal conductivity of crystals is 10–50 times lower than that of mild steel or even of stainless steel.

We will show that crusting can be lessened if we use:

– a metal with little affinity for crystals, that is, for which the interfacial adherence energy is low;

– a polished metal surface.

The superficial interfacial energy between the crystal and the metal is:

$$\gamma_{CM} = \gamma_C + \gamma_M - \beta$$

where β is the adherence energy of the crystal on the metal. γ_C and γ_M are the superficial energies of these two solids.

The interfacial energy between the metal and the mother liquor is:

$$\gamma_{ML} = \gamma_M - \gamma_L \cos\alpha_M$$

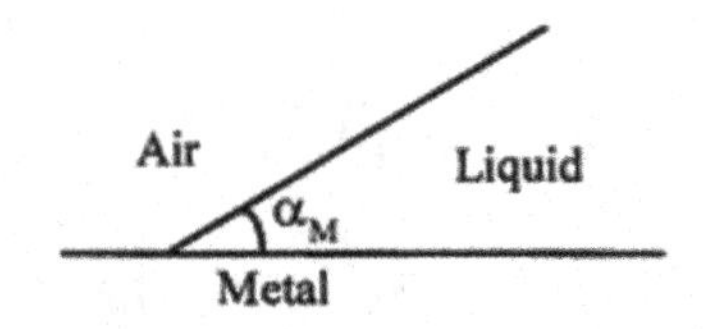

Figure 2.1. *Contact angle of the wall with the mother liquor*

The interfacial energy between a crystal and its mother liquor is:

$$\gamma_{CL} = \gamma_C - \gamma_L \cos\alpha_C = \gamma_C - \gamma_L$$

Angle α_C is zero as a crystal is always in complete contact with its mother liquor.

Now, let us consider a geometrical form representing the crystal that we suppose will grow while remaining similar to itself, where ℓ is a dimension defining crystal size.

The exterior surface of this is $\ell^2 k_\sigma$. A constant proportion σ of this surface represents contact with the solid. Proportion $(1-\sigma)$ represents contact with the liquor.

The crystal's volume is $\ell^3 k_\omega$.

The free enthalpy variation of the solid–crystal–liquor system due to the apparition of the crystal is:

$$\Delta G = \ell^2 k_\sigma \left[\sigma(\gamma_{CM} - \gamma_L) + (1-\sigma)\gamma_{CL}\right] - \frac{\ell^3 k_\omega}{\Omega}\Delta\mu$$

Ω : molar volume of the crystal.

Now, let us cancel the derivation of ΔG relative to ℓ and simplify it by ℓ.

We obtain:

$$2k_\sigma\left[\sigma(\gamma_M + \gamma_C - \beta - \gamma_M + \gamma_L \cos\alpha) + (1-\sigma)(\gamma_C - \gamma_L)\right] - 3\ell k_\omega \frac{\Delta\mu}{\Omega} = 0$$

Assuming:

$$A = 2k_\sigma\left[(\gamma_C - \sigma\beta) + \gamma_L(\sigma\cos\alpha_M + \sigma - 1)\right]$$

$$B = 3k_\omega \Delta\mu/\Omega$$

Disregarding γ_L in favor of γ_C and β, the critical size of the crystal is:

$$\ell^* = \frac{A}{B} = \frac{2}{3} \times \frac{k_\sigma}{k_\omega} \times \frac{(\gamma_C - \sigma\beta)\Omega}{\Delta\mu}$$

Crystallization will be hindered if ℓ^* is as high as possible, which means that the product $\sigma\beta$ is at a minimum. In other words:

– the minimal adherence energy β means low affinity between the metal and the crystals,

– a low value of σ implies that *the metal surface is polished.*

2.1.3. *Secondary nucleation*

For the most part, crystallizers are homogenized, which is to say that they are stirred.

During this agitation, crystals are subjected to:

– friction with the mother liquor, detaching large agglomerates or micro-particles of crystal from their surface, which also act as crystal seeds. Nevertheless, this friction is negligible for crystals of size below 4 or 5 mm, since the liquor velocity relative to the particle is very low for these sizes. As the size increases, the influence of the fluid becomes more apparent;

– impacts of the rotor on the crystals, which lead to splinters breaking off, becoming additional crystal seeds;

– collisions of crystals with each other. Clearly, this effect increases the magma content of a solid phase. We should note that a crystallizer can often function with 10–20% volume of solid material.

These three effects are a direct function of the supplementary mass agitation power (Watts per kilogram of suspension) due to the presence of crystals.

The residence time has a direct bearing on the crystal size and consequently their sensitivity to friction and impacts.

The development of an agitator system (agitator as marine propeller, blades or draft tube with recirculation propeller) is also significant.

The higher the temperature, the lower the viscosity of the liquor, which benefits the relative movements and impacts of crystals between one another or with the agitator.

Particles with sharp edges and points, that is, angular particles, are more sensitive to attrition. After abrasion, such particles take a more rounded aspect with greater resistance to abrasion.

Typically, fragments are less than 100 µm in size.

Nonetheless, impacts with the rotors of agitators or pumps can produce fragments of greater size if the impact velocity exceeds $10\,m.s^{-1}$. Indeed, in such cases, the crystals disintegrate entirely. Consequently, in order to reduce attrition, it is advisable to reduce the peripheral velocity of agitator rotors to a value less than $10\,m.s^{-1}$. However, this limitation is harder to obtain with pumps than with agitators.

For fragments smaller than 100 µm, growth G is widely dispersed, varying between 0 and $0.15.10^{-7}\,m.s^{-1}$ according to the tests performed by Wang *et al.* [WAN 92] on potassium nitrate crystals. Indeed, relative to the integration of molecules in the crystal, which can be called the effective supersaturation of the mother liquor, the usual supersaturation $\Delta\mu_0$ is different

$$\Delta\mu_{eff} = \Delta\mu_0 + \Delta\mu_{el}$$

$\Delta\mu_{el}$ is connected to the elastic stresses in the crystal and is negative.

Pohlisch [POH 87] demonstrated and verified by experimentation that the contraction (negative growth) of a crystal population due to attrition is proportional to the square of crystals' size:

$$G_a = -\frac{dL}{d\tau} \sim L^2$$

Therefore, the real growth is:

$$G = G_0 - G_a$$

Since it is typically the larger crystals that are abraded, the maximum size cannot be exceeded.

Often, in order to obtain large crystals, we must avoid accidental primary nucleation and limit the supersaturation:

$$\sigma < 0.1$$

In this case, attrition is the main source of seeds.

For low supersaturation of this order, the useful fragments (with a size greater than 20 μm and, consequently, with sufficient enlarging) represent only 1% of the total population.

If we accept the following empirical law for primary nucleation:

$$J_1 = k_N \Delta c^n$$

The secondary nucleation will be:

$$J_2 = k_{N_2} \Delta c^n$$

Moreover, in total:

$$J_1 = J_1 + J_2 = (k_N + k_{N_2})\Delta c^n$$

2.1.4. *Metastable zone: supersaturation established slowly*

Establishing the supersaturation slowly means over-cooling the solution beyond saturation at velocity v such that [NYV 85]:

$$2°C.h^{-1} < v < 20°C.h^{-1}$$

Supersaturation accessible prior to nucleation can be represented by the interval separating the saturation curve and the nucleation curve as seen in the temperature–concentration system of coordinates.

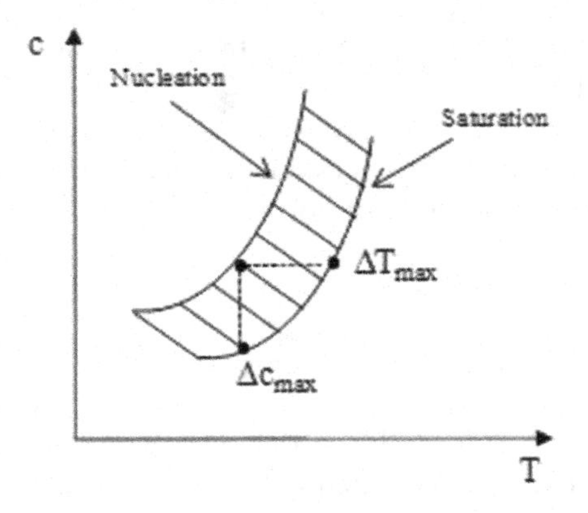

Figure 2.2. *Metastable zone*

Massive nucleation occurs once the metastability threshold is crossed. However, often ΔT_{max} and Δc_{max} do not correspond to the same nucleation curve. In Figure 2.2, the hatched area is the metastable region (labile).

2.1.5. *Measurement of nucleation order n*

The empirical laws of nucleation and growth adopted by Nyvlt *et al.* [NYV 85] are, respectively:

$$J = k_N (\Delta c)^n \qquad \text{and} \qquad \frac{dL}{d\tau} = k_G \Delta c^g$$

As supersaturation Δc increases consistently with time we will assume, like these authors, that:

$$p = \frac{dc^*}{dT} \qquad \text{and} \qquad v = \left|\frac{dT}{dt}\right| \qquad (dT/dt < 0 \text{ for cooling})$$

Accordingly, if t is the duration of time that has passed since the start of the experiment, then:

$$\Delta c(t) = c_o^* - c^*(t) \qquad \text{and} \qquad \frac{d(\Delta c)}{dt} = -\frac{dc^*}{dt} = \frac{dc^*}{dT} \times \left|\frac{dT}{dt}\right| = p\,v$$

The concentration of liquor is not sensitive to the precipitation of a few seeds, with supersaturation $\Delta c(t)$ not dependent on $c^*(t)$. Initially, $\Delta c(o) = 0$. Consequently:

$$\Delta c = p\,v\,t$$

Hence:

$$J = k_N (p\,v\,t)^n \qquad \text{and} \qquad \frac{dL}{d\tau} = k_G (p\,v\,t)^g$$

The size of a crystal formed at time τ that has grown until time τ_c at which the seeds become visible, is:

$$L = \int_{\tau}^{\tau_c} k_G \Delta c^g dt = \int_{\tau}^{\tau_c} k_G (p\,v\,t)^g dt = k_G \frac{(p\,v)^g (\tau_c^{g+1} - \tau^{g+1})}{g+1}$$

The total mass of the crystals precipitated at time τ_c is:

$$M = \int_o^{\tau_c} J(t) L^3(t) \alpha \rho_c dt = \frac{\alpha \rho_c k_G^3 k_N}{(g+1)^3} (\rho v)^{3g+n} \int_o^{\tau_c} \left(\tau_c^{g+1} - \tau^{g+1}\right)^3 t^n dt$$

Each crystal formed at time t and grew until time τ_c. Writing out:

$$I = \int_{o}^{\tau_c} (\tau_c^{g+1} - t^{g+1})^3 t^n dt = F(g,n)\tau_c^{3g+n+4}$$

With:

$$F(g,n) = \frac{3}{(n+1)(n+2+g)} + \frac{3}{2g+n+3} - \frac{1}{3g+n+4}$$

Time τ_c has passed while the difference in temperature from saturation reached value ΔT_{max}. Therefore, the moment at which the first crystals become visible is:

$$\tau_c = \Delta T_{max}/v$$

The precipitated mass is:

$$M = \frac{\alpha\rho_c k_N k_G^3}{(g+1)^3} F(g,n)(p\,v)^{3g+n}(\Delta T_{max}/v)^{3g+n+4}$$

Writing out:

$$K = \frac{\alpha\rho_c k_N k_G^3}{(g+1)^3} \times F(g,n)$$

We obtain:

$$M = Kp^{3g+n} v^{-4} \Delta T_{max}^{3g+n+4}$$

Moving to logarithms:

$$\text{Ln } M = \text{Ln } K + (3g+n)\text{Ln } p - 4\text{Ln } v + (3g+n+4)\text{Ln } \Delta T_{max}$$

Writing out:

$$m = 3g+n+4$$

We obtain:

$$\text{Ln}\Delta T_{max} = \frac{1}{m}\text{Ln M} - \frac{1}{m}\text{Ln K} - \frac{m-4}{m}\text{Ln p} + \frac{4}{m}\text{Ln v}$$

By varying v and measuring M and ΔT_{max}, we can deduce m since K and p are already known. Finally, the true order n of the empirical nucleation relation can be deduced. The measurement of M is made by passing the whole solution through a Coulter counter and not neglecting Ln M.

$$n = m - 3g - 4$$

Unfortunately, since this is a delicate process, today we prefer to deduce nucleation according to the theory of a homogenous and continuous crystallizer.

2.1.6. *Parameters influencing* ΔT_{max}

The metastable zone width is sensitive to several factors.

1) Heterogeneous nucleation:

We can compare the value of homogeneous nucleation r^* with the value of heterogeneous nucleation $r^* = \ell^*/2$ for a spherical nucleus, of which fraction σ of the surface is in contact with a solid material (ceramic or metal). Accordingly, we take $k_\omega = \pi/6$ and $k_\sigma = \pi$

$$r^*_{homogeneous} = \frac{2\gamma_c\Omega}{\Delta\mu}$$

$$r^*_{heterogeneous} = \frac{2(\gamma_c - \sigma\beta)\Omega}{\Delta\mu}$$

We observe that the critical size for heterogeneous nucleation is lower. In other words, the impurities in the suspension benefit nucleation. For an impeccably pure solution, ΔT_{max} is between 10 and 20°C, while for an untreated solution, ΔT_{max} does not exceed 2–3°C.

2) Concentration fluctuations:

These fluctuations lead to the apparition of crystal embryos of critical size. They are encouraged by:

– low viscosity;

– mechanical action (collisions, ultrasound);

– high concentration c^* at equilibrium.

In these conditions, ΔT_{max} decreases.

3) The chemical combination model:

We can consider the formation of critical cores as the end result of a series of reactions involving crystal embryos E_i containing i molecules.

$$E_1 + E_1 \rightarrow E_2$$

$$E_2 + E_1 \rightarrow E_3$$

$$E_i + E_1 \rightarrow E_{i+1}$$

$$E_{n-1} + E_1 \rightarrow E_n = E^*$$

We understand that this series of reactions is hindered:

– by the addition to the solution of an inorganic compound complexing which blocks crystallizable molecules. Thus, polyphosphates hinder the scaling of boiler tubes;

– if, for crystallization, each molecule must carry with it a high number of hydration water molecules (more generally, solvation molecules);

– if the solution is subjected to prior heating over a period of one, two or three hours at a temperature greater than the saturation temperature by 10–60°C. Essentially, this shifts the precedent reactions to the left.

Thus, in these three situations, ΔT_{max} increases.

2.1.7. *Practical study of the metastable zone (cooling)*

Let us consider the typical case in which solubility increases with temperature.

Let T^* be the saturation temperature of a solution of a given concentration. If we cool the solution from temperature T_o equal to or greater than T^*, we can make several observations:

– The developing crystals only become detectable at a temperature T_d below T^*. This temperature is lower in proportion to the cooling rate $dT/d\tau$, which is higher. This difference $\left(T^* - T_d\right) = \Delta T_{max}$ is called "metastable zone width".

– If cooling is stopped at an intermediate temperature T_i between T^* and T_d and if we then wait, after a certain delay, we observe that the crystals become detectable. This time is longer in proportion to the proximity of temperature T_o to T^*, being infinite for $T_i = T^*$. Time t_{BC} increases if the cooling is rapid for a given T_i.

When the solution has been maintained previously at temperature T_o greater than T^* for time t_0, we observe that ΔT_{max} and t_{BC} are greater in proportion to the higher values of T_o and t_o.

If we subject the liquor to mechanical stress (agitation, ultrasound, etc.), we observe that ΔT_{max} and t_{BC} decrease and can even approach zero.

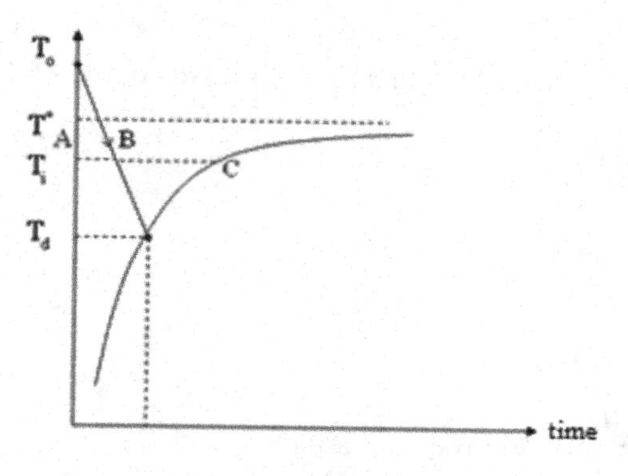

Figure 2.3. *Latency*

In all solutions, there are more or less ordered clusters (crystal embryos) of solute molecules whose size decreases as T_o increases. On cooling, this size increases slowly with time and, when it reaches a critical value, nucleation occurs, that is, germs that grow normally appear.

For given operational conditions, the metastable zone width:

– is proportional to the molar volume of the crystallized space (complex molecule, multivalent ions, high number of crystallization water molecules).

– is inversely proportional to the latent molar crystallization heat (a mean of 400 MJ per kilomole).

– increases when the crystal net has few symmetries.

– increases with the solution's viscosity.

Example 2.2.–

For Glauver salt, Nyvlt measured the metastable zone width as:

$v(°C/h)$	$\Delta T_{max}(°C)$
2	0.29
20	0.64

Hence:

$$\Delta T_{max} = 0.23 v^{0.342}$$

For slurry leaving an exchanger in which it was cooled by 0.5°C in 3 s:

$$v = \frac{0.5 \times 3600}{3} = 600°C/h$$

hence:

$$\Delta T_{max} = 2°C > 0.5°C$$

There is no nucleation in the exchanger.

However, this calculation is only indicative as cooling in exchangers occurs much faster than that in Nyvlt's tests.

2.1.8. *Interpretation of latency time*

In 1991, Dirksen *et al.* [DIR 91] proposed an expression for latency time:

$$\tau = \frac{6d^2n^*}{D\mathrm{Ln}\left(a/a^*\right)} = \frac{6d^2n^*}{D\mathrm{Ln}\left(1+\sigma\right)}$$

With:

$$n^* = \frac{4\pi r^{*3}}{3\Omega} \quad r^* = \frac{2\gamma\Omega}{kT\sigma} \quad D = \frac{kT}{3\pi\mu d}$$

γ: superficial crystal energy ($J.m^{-2}$)

Ω: molecular volume ($m^3.molecule^{-1}$)

σ: supersaturation

r^*: critical radius (of the nucleus) (m)

d: molecular diameter (m)

μ: solution viscosity (Pa.s).

This expression accounts for latency times in the order of microseconds in a liquid such as water at 20°C. In order to obtain latency of an hour for a highly saturated sucrose solution, which would be highly viscous, we cannot use the expression of D above.

Admitting that systematically we start from an initial temperature of 2–3°C above the saturation temperature, the latency time expression must account for the cooling rate. This is what we will consider next.

To begin, let us suppose that the supersaturation of the solution is obtained in a very short time relative to the latency time. This supersaturation then remains constant, which means that the crystals' nucleation and growth are typically constant through latency; in other words, crystals appear with a mass low enough not to disturb the supersaturation.

If τ is the instant at which a seed appears, the quantity of crystals that appear through time $d\tau$ and per unit volume of suspension:

$$dN = J_0 d\tau$$

These crystals grow and, admitting that the nuclei have zero diameter, the radius of a nucleus at time t is:

$$r = R(t - \tau)$$

This corresponds to volume:

$$v_g = \frac{4\pi}{3} R^3 (t-\tau)^3 = 4{,}19 R^3 (t-\tau)^3$$

The solid volume at instant t and per cubic meter of the solution is:

$$V_s = \int_0^t v_g dN = 4{,}19 J_0 R^3 \int_0^t (t-\tau)^3 d\tau = 4{,}19 J_0 R^3 \frac{t^4}{6}$$

If we accept that the solid phase becomes visible for:

$$10^{-6} < V_s = V_{vis} < 10^{-4}$$

The corresponding instant characterizes latency, which is:

$$t = \left[\frac{6 V_{vis}}{4{,}19 J_0 R^3} \right]^{1/4}$$

EXAMPLE.–

$R = 10^{-8} m.s^{-1}$ $J_0 = 4{,}7.10^{13} m^{-3} s^{-1}$ $V_{vis} = 10^{-5}$

$$t = \left[\frac{6.10^{-5}}{4{,}19 \text{ x } 4{,}7.10^{13} \text{ x } 10^{-24}} \right]^{0{,}25}$$

$$t = 23.5 \text{ s}$$

Now, supposing that we cool the solution at a constant rate dT/dt = cste, supersaturation increases proportionally to time:

$$\sigma = \frac{1}{c_0^*}\frac{dc^*}{dT}\frac{dT}{dt}t = kt$$

We accept that the equilibrium curve dc^*/dT remains constant.

The number of seeds formed per cubic meter and through time $d\tau$ is:

$$dN = J_0 d\tau$$

At time t, the number of crystals is:

$$N = \int_0^t J(\tau)d\tau$$

J depends on the instantaneous supersaturation:

$$J\left[\sigma(\tau)\right] \sim \sigma \exp\left(-\frac{A}{\sigma^2}\right) \quad \text{with} \quad \sigma = k\tau$$

The seeds that have appeared at time τ grow over time $t-\tau$. The growth equation (for a seed supposed to be spherical with radius r) is:

$$dr = R(t)dt$$

For example, if we accept the mechanism of spiral dislocations:

$$R \sim \frac{\sigma^2}{\sigma_i}\text{th}\left(\frac{\sigma_i}{\sigma}\right)$$

If we accept the mechanism by simple diffusion:

$$R \sim \sigma$$

Hence:

$$r(t,\tau) = \int_{\tau}^{t} dr = \int_{\tau}^{t} Rdt$$

The volume of a seed at time t is:

$$v_g = \frac{4\pi}{3} r^3 = v_g(t,\tau)$$

The volume of the solid phase per cubic meter of the solution is, at instant t:

$$V_s = \int_{0}^{t} J_0(\tau)\, v_g(t-\tau)\, d\tau$$

As before, the crystals become detectable when:

$$V_s = V_{vis}$$

Nielsen *et al.* [NIE 71] adopted this process, but by directly measuring J_0. Furthermore, they supposed that the growth rate R(t) was determined by crossing the diffusion layer (see sections 2.3.2 and 2.3.4). They found a time in the order of seconds.

Thus, with reference to Figure 2.3,

– development during cooling corresponds to t_{AB}

– development with a given supersaturation corresponds to time t_{BC}

The "complete latency" is:

$$t_l = t_{AB} + t_{BC}$$

2.1.9. *Nucleation by vaporization: a practical aspect*

In crystallizers that operate by vaporization, supersaturation is obtained by vaporizing the solvent. The slurry is heated in an exchanger under hydrostatic pressure, which avoids boiling in the tubes. The hot slurry is then brought up by pumping and, as the hydrostatic pressure decreases, boiling occurs throughout the ascent. An "emulsion" of small bubbles in the slurry thus enters the vapor separator located on the upper side of the installation.

Seed formation then occurs according to two distinct mechanisms:

– evacuation of water by vaporization leads to supersaturation and, consequently, classic primary nucleation.

– in return, the multiplicity of solid particles encourages the apparition of numerous, small bubbles. On the wall of an expanding bubble, we approach the dry point for the liquid solution and, in this manner, a significant quantity of bubble seeds form.

2.1.10. *Nucleation in a crystallizer (calculation)*

We can obtain the nucleation rate from population density in the following manner:

1) Homogenous cooled crystallizer:

If we accept that the nuclei are zero in size, the instantaneous nucleation rate is written as:

$$J = \left.\frac{dN}{d\tau}\right|_{L=0} = \left.\frac{dL}{d\tau}\right|_{L=0} \times \left.\frac{dN}{dL}\right|_{L=0} = G_{L=0} \times n_0$$

where N is the number of crystals per cubic meter of slurry. N_0 is the ordinate at the origin of the population density curve announcing the SDP of the crystals present in the capacity. *This expression is applicable irrespective of the value n(L).*

2) Continuous crystallizer, not necessarily homogenous:

An installation can include a number of several capacities of volume V_i in each of which nucleation occurs at rate J_i. The total number of seeds produced per unit of time is $\Sigma J_i V_i$. We can define a mean nucleation J by the relationship:

$$J \Sigma V_i = \Sigma J_i V_i$$

If we accept that all of the seeds that have appeared occur in the form of crystals in the slurry that leaves the installation at flow rate Q_0, we can write out:

$$J \Sigma V_i = Q_o N = \left[\int_0^{\infty} n(L) dL \right] \times Q_0$$

From this equality, we can deduce the value of J. We should note that N and n(L) characterize the exit product and are not indicative of the installation's mean internal granulometry.

2.2. Energetic theory of face growth

2.2.1. *Hartmann and Bennema's theory (1980)*

Experience teaches us that a crystal can usually be considered as a set of continuous pyramids, each with a shared summit that is known as the initial point. The bases of these pyramids are the crystal faces.

The surface of each base can be written as:

$$A_i = \alpha_i h_i^2$$

h_i : pyramid height

α_i : geometrical coefficient characteristic of index i pyramid.

The apparition of a crystal corresponds with the search for the minimum molecular attachment energy. Indeed, on a crystal face, the energy binding molecules to the crystal below is the attachment energy. The total of these attachment energies is consequently:

$$E_{\Sigma fix} = \sum_i \alpha_i h_i^2 E_{ifix}$$

Furthermore, the crystal volume is:

$$\Omega = \frac{1}{3} \sum_i A_i h_i = \frac{1}{3} \sum_i \alpha_i h_i^3$$

The aim is to identify a relationship that minimizes $E_{\sum fix}$, while keeping the crystal volume Ω constant. This problem is typical of those for which we apply the Lagrange multiplier method. This method implies adding the differentials of $E_{\sum fix}$ and Ω, having multiplied the latter by the coefficient λ. This total must be zero, which will imply the invalidity of d ($E_{\sum fix}$) and thereby the presence of an extremum for this function.

$$\sum_i \alpha_i h_i^2 \left(E_{ifix} - \frac{\lambda}{3} h_i \right) = 0 \qquad \text{that is,} \quad h_i = \frac{3}{\lambda} E_{ifix}$$

Accordingly, height h_i is proportional to orthogonal growth rate R_i of face i. Finally:

$$R_i = kE_{ifix}$$

Growth rate is proportional to the attachment energy of the face. This relationship was verified experimentally by Hartman and Bennena in 1980 [HAR 80]. However, this theory only concerns growth from vapor at low pressure.

2.2.2. *The presence of a solvent*

The attachment energy must be corrected by quantity ΔE_{fix} to account for the influence of the solvent that can be more or less adsorbed on the crystal [TER 01].

$$E_{fix} = E_{fix,vide} - \Delta E_{fix,sol} \qquad (\Delta E_{fix,sol} > 0)$$

2.2.3. *Morphological importance and cleavage*

Frequently (but this is not an absolute rule), the depth of the minimum interaction potential between two molecules is a decreasing function of the distance separating the two molecules when this minimum is reached. Figure 2.3 is a schematic of this.

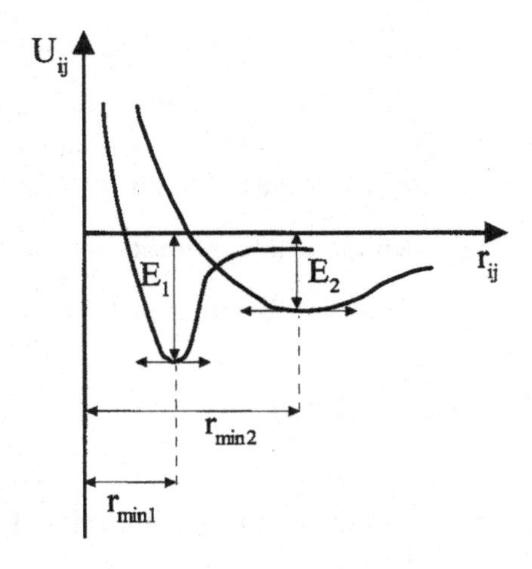

Figure 2.3. *Aspect of the minimum potential according to the intermolecular distance*

In other words:

$$\text{if } r_{min1} < r_{min2} \quad \text{so} \quad E_1 > E_2$$

We can apply the same logic to write out:

$$r_{min} = d_{hk\ell} \quad \text{and} \quad E = E_{fix}$$

This is to say that, if the interplanar distance $d_{hk\ell}$ corresponding to the direction of Miller's indices h, k, ℓ decreases, then the attachment energy increases.

$$\text{If } d_{hk\ell} \text{ decreases, then } E_{fix} \text{ increases} \qquad [2.1]$$

Of course, the easy cleavage planes are those for which E_{fix} is weak, that is for which $d_{hk\ell}$ is high.

Indeed, we have seen that (see section 2.2.1):

If E_{fix} increases, then the growth rate R increases.

However, according to relationship [2.1]:

If $d_{hk\ell}$ decreases, then R increases

Moreover, inversely:

If $d_{hk\ell}$ increases, then R decreases and the morphological importance MI increases.

The morphological importance of a face is consequently an increasing function of the interplanar distance. However, we have seen that the greater the value of $d_{hk\ell}$, the easier the occurrence of cleavage.

In conclusion, *planes of high morphological importance are planes of easy cleavage* [BRA 66].

2.2.4. *Energy aspect and kinetic perspective*

The attachment energy calculation for a layer is performed at the atomic level only by highly specialized research departments. Moreover, Hartmann and Bennema's theory [HAR 80] only concerns crystallization obtained from a low-pressure vapor. For this reason, we will now consider the kinetic growth theories that provide analytical explanations for growth R and whose parameters can be readily adapted according to experimental results.

2.3. Kinetic growth theories

2.3.1 *General*

We will now review the various models that have been proposed for face growth. We will successively study:

– The diffusion layer surrounding the crystal. This corresponds to a boundary film that must be crossed by molecules before reaching the crystal.

– The growth of K faces.

As regards to the F faces, we will examine two models:

– The spiral growth model from a dislocation that itself is a spiral.

– The "formation and spreading" model that is an improvement on the polynuclear model.

The Chernov model [CHE 61] is a combination of the spiral growth model and the model of diffusion through a diffusion layer. While not examining this model, we will proceed in another manner, following a method that is applicable both to the spiral growth model and to the "formation and spreading" model.

2.3.2. *Diffusion layer theory*

We assume that the crystal is surrounded by two superimposed molecular layers. The concentration at the interface of two layers is c_0: the external layer is the diffusion layer and the internal layer is the integration layer.

1) The diffusion layer is of thickness δ given by:

$$\delta = \frac{d}{2 + 0.6\,\mathrm{Re}^{1/2}\,\mathrm{Sc}^{1/3}}$$

d: crystal dimension (m)

Sc: Schmidt number

$$\mathrm{Sc} = \frac{\mu}{\rho D}$$

μ: liquid viscosity (Pa.s)

D: solute diffusivity in the liquid ($m^2.s^{-1}$)

ρ: liquid density ($kg.m^{-3}$)

Re: Reynolds number

$$\mathrm{Re} = \frac{u d \rho}{\mu}$$

u: velocity of the crystal relative to liquid ($m.s^{-1}$)

In a fluidized bed:

$$\mathrm{Re} = \frac{\mathrm{U d \rho}}{\varepsilon \mu}$$

ε : porosity of the fluidized bed

U : velocity of the liquid in the empty vat ($\mathrm{m.s^{-1}}$)

For an agitated suspension:

$$\mathrm{Re} = \mathrm{d}^{4/3}\mathrm{W}^{1/3}\rho\mu^{-1}$$

W: mechanical power per unit of suspension mass ($\mathrm{W.kg^{-1}}$).

Molar concentration ($\mathrm{kmol.m^{-3}}$) on the internal face of the diffusion layer is c_0. On the external face, it is equal to c_∞ which is the concentration inside the solution.

The molar flow density through the diffusion layer is:

$$\varphi = \frac{D}{\delta}(c_\infty - c_0) \qquad (\mathrm{kmol.m^{-2}.s^{-1}})$$

2.3.3. *Integration mechanisms*

A face's growth varies according to its nature.

K faces have extremely fast growth since, as soon as a molecule comes into contact with the surface, it is effectively fixed to its point of impact.

But only faces that develop slowly have morphological significance. These faces are the F faces.

The K faces are considered to have a continuous growth as opposed to the F faces that develop by successive layers. The edge of a layer is a step (like the step of a stair). Steps develop by successive lines.

If, for the sake of simplicity, we think of the molecules as cubes, these cubes preferentially bind to the edge of a step where a cavity is present, that is, at the place where the cube is held by three faces. This location is known as a “cavity”.

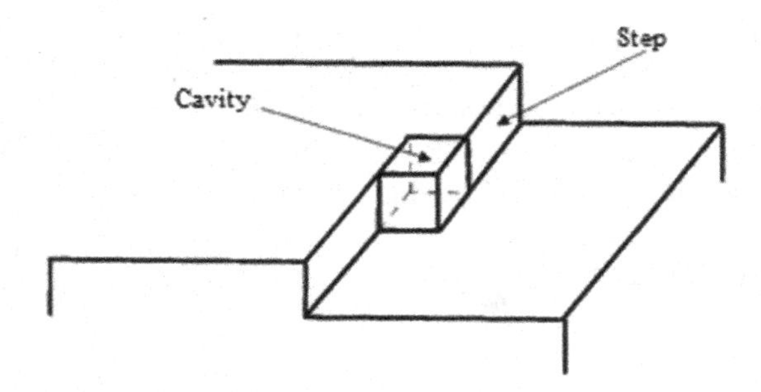

Figure 2.4. *Binding in a cavity on the edge of a step*

Cavities on the edge of a step are a relatively frequent occurrence. On the other hand, steps only occur from spiral dislocations.

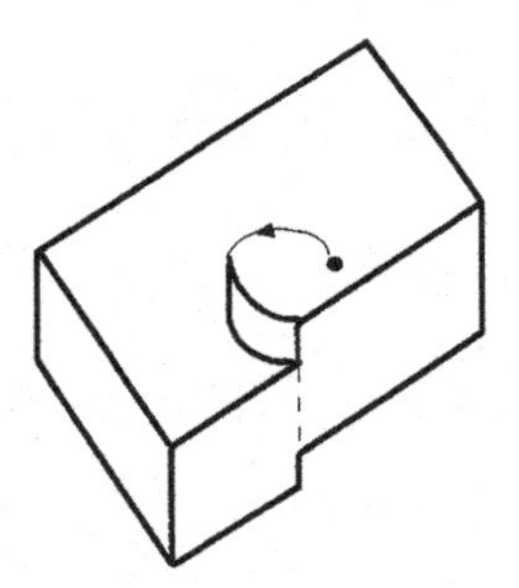

Figure 2.5. *Spiral step*

These mechanisms give rise to faces that are microscopically flat. However, if the steps’ linear energy decreases, growth occurs by formation (bidimensional nucleation) and spreading with the formation of mounds in the shape of stepped pyramids. As each mound spreads due to its horizontal surface increasing, a new bidimensional nucleus forms at its summit, which then in turn spreads itself.

Figure 2.6. *Stepped mound*

Growth by spiral dislocations results in smooth surfaces, while the mechanism of "formation and spreading" leads to rough surfaces.

2.3.4. *Fundamental parameters*

The mean attachment energies are [BUR 51]:

$$\varphi \# 4kT$$

Hence, the activation energy for diffusion:

$$\frac{\Delta G_{diff}}{kT} = \frac{6\,to\,8\varphi}{20kT} \# 0.35 \frac{\varphi}{kT} = 1.4$$

From this, we can deduce the diffusivity on the surface:

$$D_s = a^2 \nu e^{-1,4}$$

a : molecule dimension (part of a multilayer) (m)

$$2.10^{-10}m < a < 10.10^{-10}m$$

ν: vibration frequency of an atom or a molecule in position (s^{-1})

Frequency ν increases with molecule complexity:

$$10^{13}s^{-1} < \nu < 10^{15}s^{-1}$$

The characteristic length for surface diffusion is:

$$\lambda_s = a \exp\left(\frac{W}{4kT}\right) \quad \text{with} \quad W = 6\,to\,8\varphi$$

Hence:

$$\lambda_s \# a\, e^7$$

The activation energy ΔG_{niche} is equal to 1/4 of the energy 2φ required for moving a molecule from a position on this step towards a cavity on the same step.

$$\frac{\Delta G_{cavity}}{kT} = \frac{1}{4} \times \frac{2\varphi}{kT} = 2$$

NOTE.–

Surface energy.

Through the following, the superficial energy of a crystal is:

$$10^{-3}\,\text{J.m}^{-2} < \gamma < 0.1\,\text{J.m}^{-2}$$

The energy use per unit of step length (edge unit) is:

$$2.10^{-12}\,\text{J.m}^{-1} < \gamma_\ell < 10^{-10}\,\text{J.m}^{-1}$$

Indeed, $\gamma_\ell = \gamma a$, where a is the dimension of one molecule. *We can refer to the publication of Nielsen et al. [NIE 71] for the various values of* γ.

2.3.5. *Growth of a K face*

$$R = h\nu\sigma \exp\left(-\frac{\Delta G_{cavity}}{kT}\right)$$

[MUT 01, p. 294].

EXAMPLE 2.3.–

$h = 5.10^{-10}\,\text{m}$ $\qquad$ $\nu = 10^{14}\,\text{s}^{-1}$ $\qquad$ $\sigma = 0.05$

$$\Delta G_{cavity}/kT = 2$$

$$R = 5.10^{-10} \times 10^{14} \times 0.05\exp(-2)$$

$$R = 338 \text{ m.s}^{-1}$$

This is an absurd result. Consequently, we will involve the diffusion layer in the following example:

EXAMPLE II.–

$d = 10^{-3}\,\text{m}$ $\quad\mu = 10^{-3}\,\text{Pa.s}$ $\quad\rho = 1100\,\text{kg.m}^{-3}$

$u = 0.005\,\text{m.s}^{-1}$ $\quad D = 10^{-9}\,\text{m}^2.\text{s}^{-1}$ $\quad\Omega c^* = 0.1$

$$\sigma = 0.05$$

$$\text{Re} = \frac{5.10^{-3} \times 10^{-3} \times 1100}{10^{-3}} = 5,5$$

$$\text{Sc} = \frac{10^{-3}}{1.1\times 10^{3} \times 10^{-9}} = 900$$

$$\delta = \frac{10^{-3}}{2 + 0.6 \times 2.34 \times 9.99} = 0.624.10^{-4}\,\text{m}$$

As the kinetics are particularly fast, we can accept that:

$$c_0 \# c^*$$

$$R = \Omega c^* \frac{D}{\delta}\left(\frac{c_0 - c^*}{c^*}\right) = \Omega c^* \frac{D}{\delta}\sigma = \frac{0.1 \times 10^{-9} \times 0.05}{0.624.10^{-4}}$$

$$R = 0.801.10^{-7}\,\text{m.s}^{-1}$$

2.3.6. *Bidimensional nucleation*

The creation free enthalpy of flat crystal embryos (round in cross-section and of radius r) from a supersaturated solution is:

$$\Delta G_2 = 2\pi r \gamma_\ell - \frac{\pi r^2 h}{\Omega} \Delta\mu$$

γ_ℓ: edge energy ($J.m^{-1}$)

h: embryo height (m)

Ω: molar volume of solute ($m^3.molecule^{-1}$).

Let us regard the derivation of ΔG_2 in order to find the maximum for this function

$$\frac{d(\Delta G_2)}{dr} = 2\pi\gamma_\ell - \frac{2\pi r h}{\Omega}\Delta\mu = 0$$

$$r^* = \frac{\gamma_\ell \Omega}{h\Delta\mu} \quad \text{hence} \quad \frac{\Delta G_2^*}{kT} = \frac{\gamma_\ell^2 \Omega}{kTh\Delta\mu} = \left(\frac{\gamma}{kT}\right)^2 \frac{\pi a^4}{\nu Ln(1+\sigma)}$$

We have added coefficient ν, which is the number of ions that correspond to the possible dissociation of an electrolyte.

2.3.7. *Spiral growth [BUR 51]*

$$R = \zeta a \nu \exp\left(-\frac{\Delta H}{kT}\right)\frac{\sigma^2}{\sigma_1} th\left(\frac{\sigma_1}{\sigma}\right)$$

$$\zeta = \left[1 + \frac{D_s}{\beta\lambda_s} th\left(\frac{\sigma_1}{\sigma}\right)\right]^{-1}$$

ΔH : molar dissociation heat ($J.molécule^{-1}$)

Let us specify the use of the exponential term here.

A physical or chemical phenomenon, and in particular crystallization, that is significantly faster when exothermic, or less endothermic (which is essentially equivalent).

Since dissolution is the inverse path of crystallization, crystallization that is somewhat endothermic corresponds to dissolution that is somewhat exothermic.

In other words, if ΔH is the positive heat released by dissolution, the crystallization rate is a decreasing function of ΔH, which effectively announces the exp factor (−ΔH/RT).

D_S: surface diffusion ($m^2.s^{-1}$)

$$\lambda_s = a\, e^7 \qquad D_s = a^2 \nu e^{-1,}$$

$$\lambda_0 = \frac{4\pi\gamma_\ell a^2}{kT\sigma} \qquad \sigma_1 = \frac{2\pi\gamma_\ell a^2}{kT\lambda_s}$$

$$\beta = \frac{a^2\nu}{\lambda_0}\exp\left(-\frac{\Delta G cavity}{kT}\right)$$

a: dimension of one molecule (part of a multilayer) (m)

e: base of natural logarithms (2.3026)

ν: fundamental vibration frequency of the molecule (s^{-1})

The vibration frequency ν decreases with the molecule size. It is in the order of 10^{14} for a monatomic molecule (see [MUT 01])

σ: relative supersaturation (see section 2.1.1)

$$\sigma = \frac{c - c^*}{c^*} cn \text{ or rather} \frac{c}{c^*} = 1 + \sigma$$

EXAMPLE 2.4.–

$\gamma_\ell = 12.10^{-12}$ $\quad$ $a = 5.10^{-10}\,m$ $\quad$ $\sigma = 0.5$ $\quad$ $\delta = 1.07.10^{-4}\,m$

$k = 1.38.10^{-23}\,J$ $\quad$ $\nu = 10^{14}\,s^{-1}$ $\quad$ $\Delta G_{cavity}/kT = 2$ $\quad$ $\Delta H/kT = 12$

$$D_s = 25.10^{-20} \times 10^{14} \times 0.246 = 6.16.10^{-6}\,m^2.s^{-1}$$

$$\lambda_s = 5.10^{-10} \times 1096 = 5.48.10^{-7}$$

$$\lambda_0 = \frac{4 \times \pi \times 12.10^{-12} \times 25.10^{-20}}{1.38.10^{-23} \times 300 \times 0.5} = 1.821.10^{-8}\,m$$

$$\beta = \frac{25.10^{-20} \times 10^{14} \times e^{-2}}{1.821.10^{-8}} = 185\,m.s^{-1}$$

$$\sigma_1 = \frac{2\pi \times 12.10^{-12} \times 25.10^{-20}}{1.38.10^{-23} \times 300 \times 5.48.10^{-7}} = 0.00831$$

$$th\left(\frac{\sigma_1}{\sigma}\right) = 0.01662$$

We note that this value differs only slightly from:

$$\frac{\sigma_1}{\sigma} = \frac{0.00831}{0.5} = 0.01662$$

$$\zeta^{-1} = 1 + \frac{6.16.10^{-6}}{185 \times 5.48.10^{-7}} \times 0.01662 = 1.001$$

$$R = \frac{1}{1.001} \times 5.10^{-10} \times 10^{14} \times e^{-12} \times \frac{(0.5)^2}{0.00831} \times 0.01662 = 0.153\,m.s^{-1}$$

This result is very high. In reality, intense nucleation multiplies the seeds, significantly decreasing supersaturation. If σ becomes equal to 0.005, we arrive at the following results:

$\lambda_0 = 1.821.10^{-6}\,m$ $\quad$ $\sigma_1/\sigma = 1.662$ $\quad$ $\zeta^{-1} = 6.6295$

$\beta = 1.858\,m.s^{-1}$ $\quad$ $th(\sigma_1/\sigma) = 0.9305$ $\quad$ $R = 0.00013\,m.s^{-1}$

The growth rate is divided by more than 1,000.

In any case, these rates are significantly greater than those resulting from crossing the diffusion layer (see section 2.3.5), which is consequently limiting.

2.3.8. *"Formation and spreading"*

$$R = hJ^{1/3}R_{2D}^{2/3}$$

The number of nuclei appearing per unit of surface is:

$$J = \frac{\nu\sigma^{1/2}}{a^2}\exp\left(-\frac{\Delta G_2^*}{kT}\right)$$

The progression rate of the steps is:

$$R_{2D} = \frac{D_s\sigma}{\lambda_s}$$

Hence, the growth rate:

$$R = h\sigma^{5/6}\left(\frac{\nu}{a^2}\right)^{1/3}\left(\frac{D_s}{\lambda_s}\right)^{1/3}\exp\left(-\frac{\Delta G_2^*}{3kT}\right)$$

EXAMPLE 2.5.–

$a = h = 5.10^{-10}\,m$ $\quad\nu = 10^{14}$ $\quad D_s = 6.16.10^{-6}\,m^2.s^{-1}$

$\sigma = 0.5$ $\quad\Delta G_2^*/kT = 70.63$

$$\lambda_s = 5.10^{-10}e^7 = 5.48.10^{-7}\,m$$

$$R = 5.10^{-10} \text{ x } 0.5^{5/6} \text{ x } \left(\frac{10^{14}}{25.10^{-20}}\right)^{1/3}\left(\frac{6.16.10^{-6}}{5.48.10^{-7}}\right)\exp\left(-\frac{70.63}{3}\right)$$

$$R = 6.18.10^{-9}\,m.s^{-1}$$

Note that if ΔG_2^* is nullified, that is, if the linear energy γ_ℓ is nullified, growth becomes:

$$R_0 = 2.57.10^{-7} \times 1.20.10^6 = 103 \text{ m.s}^{-1}$$

Growth becomes of the same order as that of a K face and is consequently limited for the transfer through the diffusion layer.

2.3.9. *Diffusion–integration combination*

According to results obtained previously, the growth rate's integration mechanisms are announced by the following equations:

Formation and spreading $R_I = D\sigma^{5/6}\exp(-B/3\text{LnS})$

Spiral dislocations $R_I = E\sigma^2\text{th}(\sigma_1/\sigma)$

All of these functions increase monotonically according to the relative supersaturation σ. They are zero for $\sigma = 0$.

The molar flow density crossing the diffusion layer is:

$$\varphi = \frac{D}{\delta}(c_\infty - c_0) = \frac{D}{\delta}c^* N_A \sigma$$

This corresponds to a growth rate of:

$$R_D = \Omega\varphi = \frac{D\Omega c^* \sigma_D}{\delta}$$

c^* : solution concentration at equilibrium (kmol.m^{-3})

Ω : molecular volume ($\text{m}^3.\text{kmol}^{-1}$).

In addition, we can write out:

$$\sigma_\infty = \frac{c_\infty - c^*}{c^*} = \frac{c_\infty - c_0}{c^*} + \frac{c_0 - c^*}{c^*} = \sigma_D + \sigma_I$$

c_0 is the concentration at the interface separating the diffusion layer and the integration layer.

The calculation of σ_I is performed by equating rate $R_I(\sigma_I)$ for integration and rate $R_D(\sigma_D)$ for diffusion.

$$R_I(\sigma_I) = R_D(\sigma_D)$$

An analytical solution is possible in both extremes of the dislocation mechanism ($\sigma_1 << \sigma$ and $\sigma_1 >> \sigma$).

The “formation and spreading” mechanism is not susceptible to a numerical resolution.

2.3.10. *Retrodiffusion of solvent*

On binding to the crystal, the solute (or, more generally, the crystallization unit) takes the place of the solvent that was in contact with the face in question. Let us write that the global volume is unchanged.

$$N_s\Omega_s + N_u\Omega_u = 0 \qquad [2.2]$$

N: flow density towards lining ($\text{kmol.m}^{-2}.\text{s}^{-1}$)

Ω: molar volume ($\text{m}^3.\text{kmol}^{-1}$).

Index “s” relates the solvent and index “u” to the crystallization units.

The Maxwell–Stéfan equation is written as:

$$-\frac{c_u}{RT}\frac{d\mu_u}{dz} = \frac{N_u x_s - N_s x_u}{D} \qquad [2.3]$$

with:

$$x_s + x_u = 1 \qquad [2.4]$$

x: molar fractions.

Using equations [2.2] and [2.4], let us replace x_s and N_s in [2.3].

We obtain:

$$-\frac{Dc_u}{RT}\frac{\partial \mu_u}{\partial z} = N_u K \quad \text{with} \quad K = 1 + x_u\left(\frac{\Omega_u - \Omega_s}{\Omega_s}\right)$$

Indeed, the Fick law can be written out:

$$-\frac{Dc_u}{RT}\frac{\partial \mu_u}{\partial z} = N_u$$

We observe that the retrodiffusion of solvent divides flow density N_u towards the crystal via crystal corrective coefficient K. However, we should note that the coefficient is quite close to 1, and moreover, can be above or below this value.

2.3.11. *Conclusions*

1) The kinetics referred to as mononuclear and polynuclear, such as those presented by Dirkson *et al.* [DIR 91], give rates that are far too low and disconnected from experiment data.

Indeed, the population density (frequency) in a continuous homogenous crystallizer is:

$$n(L) = n_0 \exp\left(-\frac{L}{G\tau}\right)$$

L: crystal dimension (m)

G: crystal growth ($m.s^{-1}$)

$$G = 2R$$

τ : residence time in the crystallizer (s)

where G is constant and Ln[n(L)] is a decreasing linear function of L (see Figure 2.7). The curve, in absolute value, increases significantly if

G decreases, which is the case of crystals at the beginning of their existence as they are not carrying dislocations and cannot begin to grow except by "formation and spreading", which is very low. This explains the rounded form of the curve $Ln[n(L)] = f(L)$ for the very low L. The transition between the rounded form and the rectilinear form occurs for a value of L between 50 and 100 µm. This is the size at which dislocations begin to appear.

NOTE.–

Ostwald maturation:

When large and small crystals are kept in a continuous phase, the small crystals disappear, benefiting the larger crystals. This phenomenon is significant in metallurgy where the continuous phase, the "matrix", is in a solid state. For the matter at hand, this phase is liquid and typically of low viscosity, so that a decantation, or more commonly, a separation, intervenes long before maturation occurs. On this question, which we will not address here, there are several works of reference value [RAT 02, MAR 84].

Nonetheless, we should note that the influence of gravity could be neutralized by placing the suspension in a cylindrical recipient that turns slowly on its horizontal axis. A good example of this would be the maturation of sugar seeds.

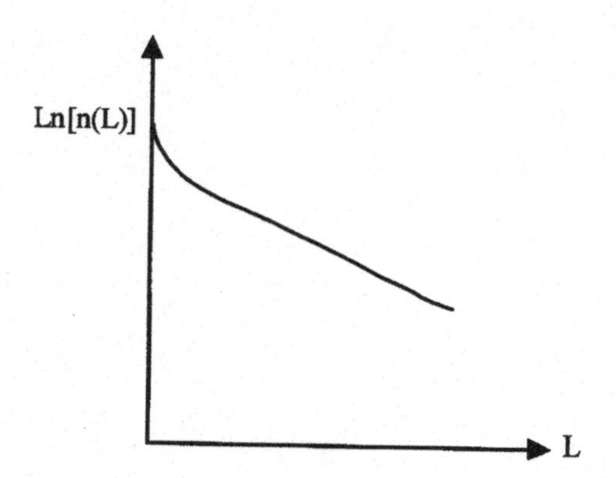

Figure 2.7. *Aspect of logarithmic variations in population density according to crystal dimension*

2.3.12. *Drawing crystal shapes [DOW 80]*

Relative to the center (starting point) of the crystal, we represent each face by the classic plane equation:

$$\frac{x}{h}+\frac{y}{k}+\frac{z}{\ell}=R\tau$$

R : face growth rate

h, k, ℓ : Miller indices

τ : time

Triplet faces are resolved in order to give the summits. The required polyhedron is the minimum polyhedron that can be obtained by eliminating the summits that are further from the center than all of the faces. The edges are obtained by connecting the summits that have two faces in common.

Rotation matrices allow for the crystal to be turned in space.

3

Crystallization in a Sugar Refinery

3.1. Theory of sugar crystallization

3.1.1. *Growth mechanisms*

The growth kinetic is consistent with Burton, Cabrera and Franck's BCF law (see section 2.3.7).

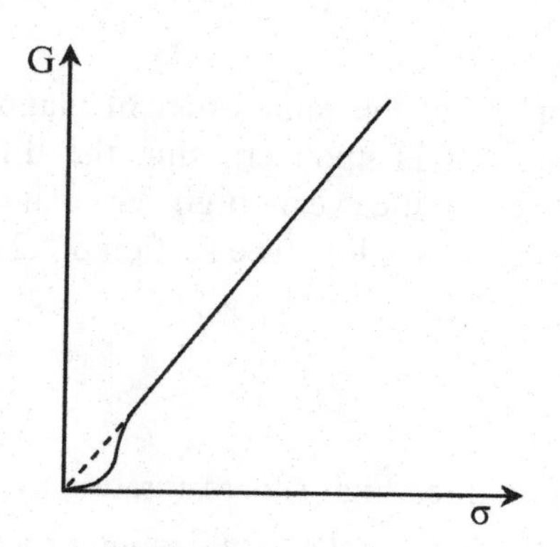

Figure 3.1. *Growth variations according to supersaturation (BCF law)*

Non-sugars are adsorbed on the surface, taking the place of the sugar molecules and slowing the integration of molecules on the crystalline faces. Accordingly, BCF growth should be multiplied by a reduction factor K between 0 and 1 and which is an increasing function of the purity (see section 3.1.3) according to Table 3.2, taken from the work of Bartens [BAR 98].

Purity	0.7	0.8	0.9	1
K	0	0.09	0.38	1

Table 3.1. *Reduction factor*

The supersaturation that we typically employ depends on the quality of sugar present in the magma. We can draw the table as follows:

Refined sugar	1.05
Sugar A (white)	1.10–1.15
Sugar B (raw)	1.10–1.20
Sugar C (after production)	1.15–1.25

Table 3.2. *Typical supersaturations* $S = 1 + \sigma$

The BCF growth rate is of the same order of magnitude as that of other crystals (organic or not). It is upon crossing the diffusion layer that the transfer is slowed, due to the very high viscosity of sugar solutions. Consequently, diffusivity is very low (see section 3.2.2).

3.1.2. *Nucleation*

Primary nucleation requires high supersaturation ($\sigma \geq 0.3$) and is difficult to control. For this reason, industrial crystallization proceeds by seeding.

Secondary nucleation is a result of the friction of liquid on the crystals' surface. This friction tears away clusters of molecules that subsequently act as seeds for secondary nucleation. Indeed, Brown *et al.* [BRO 92] demonstrated that the number of seeds created in this way is proportional to the crystalline surface present in the magma.

Liang *et al.* [LIA 87] showed that crystals smaller than 30 μm all grow at a constant rate, but that this rate is distributed according to a Gaussian law by which:

– the mean value is given by the BCF law;

– the standard deviation is given by:

$$\sigma^2 = 0.14G^{-2}$$

These crystals (below 30 μm in size) are the result of crystals being affected by the agitator or impeller.

Note that secondary nucleation is not successful if the supersaturation is greater than 0.2. This is a necessary condition for the development of agglomerates which arise near crystal faces.

In practice, we try to avoid secondary nucleation, and often:

$$\sigma < 0.2$$

3.1.3. *Sugar purity and weight ratio*

Purity is defined by:

$$q_s = \frac{\text{sugar mass}}{\text{dry material mass}}$$

Impurity is defined as:

$$q_{ns} = \frac{\text{non sugar mass}}{\text{dry material mass}}$$

Of course:

$$q_s + q_{ns} = 1$$

In solution, the weight ratio of sugar is:

$$X_s = \frac{\text{sugar mass}}{\text{water mass}}$$

For purity equal to 1, the saturation rate driven by Grut (see [DE 83]) is defined according to temperature:

$$X_s^* = 256.35/(147.47 - t) \qquad (t{:}^\circ C)$$

Where there are impurities present on saturation, we write:

$$X_s^* = X^* \, y_{sat}$$

According to Wagnerowski [WAG 62]:

$$y_{sat} = 0.178X_{ns} + 0.820 + 0.18\exp(-2.1X_{ns})$$

with:

$$X_{ns} = \frac{\text{impurity mass}}{\text{water mass}}$$

3.2. Orders of magnitude

3.2.1. *Boiling delay*

This delay depends somewhat on the liquid-phase temperature. The following table provides the values according to the weight ratio X of sugar in the liquid. A more complete treatment is given by Bubnik *et al.* [BUB 95].

X%	35	50	60	70	80	90	(X : weight ratio)
R_E°C	1	2	3	5	9	17	

Table 3.3. *Boiling delay RE*

3.2.2. *Diffusivity*

Austmeyer [AUS 81] proposed an algorithm and curve network from which we have taken the following approximate values (in 10^{-12} $m^2.s^{-1}$).

X % \ t°C	50	70	90
30	0.5	0.75	1.1
60	0.2	0.30	0.40
70	0.15	0.20	0.25

Table 3.4. *Diffusivity*

Further details are available in Austmeyer's thesis [AUS 81] or in the work of Bartens [BAR 98].

3.2.3. *Viscosity*

3.2.3.1. *Liquid viscosity*

This applies for:

– the transfer of matter through the diffusion layer;

– the calculation of thermal transfer;

– centrifugal filtration (this can be divided by 2 per reheating).

Purity influences the liquid's viscosity, diminishing it by 10–20% when it goes from 100 to 60%.

Schliephake *et al.* [SCH 83] provided a network of curves for the viscosity of pure solutions and technical solutions according to the weight ratio in dry material, temperature and purity.

Referring to the table of approximate values relative to T and of M.S.%, we see that purity has some influence:

M.S.% \ °C	40	60	80	[SCH 83]
40	3.5	2	1	
60	25	10	5.5	
80	1 050	300	80	

Table 3.5. *Solution viscosity (centipoises)*

3.2.3.2. *Magma viscosity*

This applies to the agitation power. At 1,500 Pa.s, the impeller rotation speed is reduced to 0.5 rev.mn^{-1}. Magma viscosity can be divided by 10 per heating or dilution.

(The crystal content and liquid viscosity are involved separately for the thermal exchange calculation.)

M.S.% \ °C	40	60	80
90	138	30	10
92	300	70	14
95	1 400	300	103

Table 3.6. *Magma apparent viscosity (Pa.s)*

3.2.4. *Properties of crystallized sugar*

Superficial energy $\gamma = 0.224$ J.m^{-2}

Density $\rho_s = 1590$ kg.m^{-3}

Molar mass $M = 342.2$ kg.kmol^{-1}

Thermal conductivity $\lambda = 0.58$ W.m^{-1}.K^{-1}

Water conductivity $\lambda_e = 0.45$ W.m^{-1}.K^{-1}

Melting temperature $T_f = 186°C$

Molar heat capacity $C_s = 425.8.10 \text{ J.kmol}^{-1}.\text{K}^{-1}$

Fusion enthalpy $\Delta H_{fus} = 46.41.10^6 \text{ J.kmol}^{-1}$

Dissolution (in water) is athermal.

3.3. Crystallization kinetics

EXAMPLE 3.1.–

Let us consider the crystallization of sugar solution in water at 40°C. We will use the BCF expression (see section 2.3.7):

$a = 10^{-9}$ m $\quad \nu = 10^{14} \text{ s}^{-1} \quad \Delta H_{diss} = 0$

$\sigma = 0.2 \quad \gamma = 0.224 \text{ N.m}^{-1} \quad \nu = 10^{14} \text{ s}^{-1}$

$T = 313.15$ K $\quad \Delta G_{nich}/kT = 2 \quad k = 1.38.10^{-23} \text{ J.molecule}^{-1}.\text{K}^{-1}$

$$D_s = 10^{-18} \times 10^{14} \times 0.2465 = 0.2465.10^{-4} \text{m}^2.\text{s}^{-1}$$

$$\lambda_s = 10^{-9} \times 1097 = 1.097.10^{-6} \text{m}$$

$$\gamma_\ell = 0.224.10^{-9} \text{J.m}^{-1}$$

$$\lambda_o = \frac{4\pi \times 0.224.10^{-9} \times 10^{-18}}{1.38.10^{-23} \times 313.15 \times 0.2} = 3.256.10^{-6} \text{m}$$

$$\beta = \frac{10^{-18} \times 10^{14}}{3.256.10^{-6}} \exp(-2) = 4.56 \text{m.s}^{-1}$$

$$\sigma_1 = \frac{2\pi \times 0.224.10^{-9} \times 10^{-18}}{1.38.10^{-23} \times 313.15 \times 1.097.10^{-6}} = 0.297$$

$$\sigma_1/\sigma = 0.297 / 0.20 = 1.48$$

$$th(\sigma_1/\sigma) = \frac{4.1653}{4.6205} = 0.9015$$

$$\zeta^{-1} = 1 + \frac{0.2465.10^{-4} \times 0.9015}{4.156 \times 1.097.10^{-6}} = 5.874$$

$$R = \frac{10^{-9} \times 10^{14} \times \exp 0 \times 0.04 \times 0.9015}{0.297 \times 5.874}$$

$$R = 2067 \text{ m.s}^{-1}$$

This rate is unrealistic, but could be reduced if the frequency chosen for the molecular vibrations were lower or if the dissolution heat were positive. In reality, we group these two values together in an empirical coefficient K.

$$K = \zeta \nu \exp\left(\frac{-\Delta H_{diss}}{RT}\right)$$

EXAMPLE 3.2.–

Let us examine the transfer through the diffusion layer (with $\sigma_{diff} = 0.2$) (see section 2.3.2):

$d = 0.7.10^{-3}$ m $D = 0.15.10^{-12}$ m^2.s^{-1} $T = 40$°C

$W = 1$ W.kg^{-1} $\rho_s = 1590$ kg.m^{-3} $\sigma = 0.2$

$$X_s^* = \frac{256.35}{147.47 - 40} = 2.385 \quad \text{(see [DE 83])}$$

On saturation, the volume fraction of sugar is:

$$\Omega c^* = \frac{2.385/1590}{\frac{1}{1000} + \frac{2.385}{1590}} = \frac{0.0015}{0.001 + 0.0015} = 0.6$$

$$\rho = \frac{1+2.385}{\frac{1}{1000}+\frac{2.385}{1590}} = \frac{3.385}{0.0025} = 1350\ \text{kg.m}^{-3}$$

$$\%\text{M.S.} = \frac{2.385}{3.385} = 70.46\ \%$$

$$\mu = 100\ \text{cp} = 0.1\ \text{Pa.s}$$

$$\begin{aligned}\text{Re} &= (0.7.10^{-3})^{1.333} \times 1 \times 1350 \times 10\\ &= 0.8391\end{aligned}$$

$$\text{Sc} = \frac{0.1}{1350 \times 0.15.10^{-12}} = 0.494.10^{9}$$

$$\delta = \frac{0.7.10^{-3}}{2+0.6 \times 0.916 \times 790} = 1.603.10^{-6}\ \text{m}$$

$$\text{R} = \frac{0.6 \times 0.15.10^{-12} \times 0.2}{1.603.10^{-6}} = 1.1.10^{-8}\ \text{m.s}^{-1}$$

3.3.1. *Integration layer and diffusion layer combination*

As the integration rate (example I) is extremely high, the crystallization rate is determined by crossing the diffusion layer (example II).

Thus, *any experimental value of R must be interpreted by taking the diffusion layer into account in order to correctly assess supersaturation.*

3.4. Practice of sugar crystallization

3.4.1. *The three crystallization techniques*

These are the only three techniques that can be used to make crystals (with the exception of chemical precipitation).

1) Vaporization of the solvent by heating with vapor to obtain sugar A occurs with a vapor pressure in the order of 0.18–0.20 bar abs for 3 h at

70°C. Of course, a preparation time of more than 20 mn at 80°C provokes crystal and liquid coloration, particularly if the invert sugar content exceeds a certain limit and if the juice is not sulfated. This is why we often proceed by cooling.

We mix the exit crystallization A syrup with the thick syrup. The solvent's vaporization then gives the sugar B and green syrup.

2) Cooling by expansion. Depending on the circumstances, cooling by 80–60°C or by 70–40°C (pressure: 0.05 bar abs) can occur. Green syrup must be added to reduce the apparent viscosity of the magma from 1,500 to 150 Pa.s.

This occurs progressively at a cooling rate of 7–11°C.h^{-1}.

Crystallization by cooling is used in the final phase, that is, for sugar C together with the production of seed magma. However, cooling can be produced not only by expansion but also by an exterior agent.

3) Cooling by water or air. The sugar A magma is cooled from 70 to 35°C at a rate of 6°C.h^{-1}, which can increase the production by 30% compared with simply steam vaporizing water. However, for sugar A, the steam vaporization is typically preferred over cooling.

Sugar B magma is cooled from 80 to 50°C with service water at a rate of 10°C.h^{-1}. With air as a cooling agent, we can only cool to 80–60°C in 10 h or 2°C.h^{-1}.

For sugar C, cooling is preferred to vaporization of water in the juice, at a rate of 1–1.5°C.h^{-1} from 80 to 40°C.

The temperature difference between water and magma is from 10 to 15°C. Cooling occurs with tubes of diameter 10–15 cm through which the magma circulates at a velocity above 0.1 m.s^{-1}. Surface integration is crucial, with viscosity not exceeding 15 Pa.s.

NOTE.– The production of steam vapor during vaporization does not present significant problems of liquid entrainment, since droplets are larger and denser than for standard crystallizations.

3.4.2. *Steps involved in industrial sugar crystallization*

The installation is supplied with thick juice of 92% purity.

A crystallizer's magma is called massecuite. By centrifugal dewatering, this provides crystallized sugar and a filtrate.

Crystallization occurs as a cascade in three steps that give three qualities of sugar: A, B and C.

	Magma M.S.	Filtrate purity	Magma purity	Crystal fraction	Crystal purity
A	89–92	80	90–95	25	> 99
B	93–94	70	88		94–98
C	95–96	60	76	55	81–93

Table 3.7. *Qualities of types of sugar*

Beetroot is only affected by sugars A and C.

The filtrate of massecuite C is known as molasses.

In reality, numerous processes exist with various means of recycling sugar or filtrate. Details can be found in the works of Bartens [BAR 98].

Step C works with a greater supersaturation than for A as the liquid-phase viscosity is greater, which reduces the diffusivity of sugar together with its transfer to the solid phase. This must be compensated by the concentration gradient, both across the diffusion film and the phase of integration to the solid.

NOTE.– A series of at least six theoretical crystallizers is required (for example, 10 real compartments in order to approach vat operation for crystals that are uniform in size).

The crystals have a mean size in the order of 0.8 mm for quality (A). This size decreases in the ratio of 2 for quality (B) and in the ratio of 3 for quality (C).

3.4.3. *Seed preparation*

The seed is not prepared by high supersaturation primary nucleation, but rather by sampling in production, followed by milling, and then maturation.

1) Sampling: sugar C, for example, is taken in order to manufacture the seed for sugar B.

2) Milling: this operation occurs in a small ball mill of around 1 m^3 over 4 h. This takes place as a wet process (isopropanol). The isopropanol leads to the formation of aggregates.

3) Maturation: the seed is kept for 4 weeks in a drum (of 1 m^3 in volume) that turns slowly, dividing the number of particles by two or three. The mean diameter of the future seeds is of the order of 10 µm, and their size distribution is relatively uniform. The slurry contains from 33 to 35% crystals by weight.

4) Addition of seed: the alcohol is vaporized and the crystals remain in suspension in the aqueous sugar solution. The crystallizer's mother liquor saturation must be in the order of 0.95–1 during the introduction of seed, which allows for the dissolution of microcrystals. Saturation then increases to the following values:

Refined sugar	1.05
White sugar (A)	1.25
Raw sugar (B)	1.20
Sugar after production (C)	1.25

A crystallizer has a draft tube and internal thermal exchange. The tubes are short (1–1.5 m in length) and of low inner diameter (1–1.5 cm). The seed volume must reach 25% of the magma volume. The useful volume is consequently reduced to 75% of the total volume.

In order to eliminate the agglomerates, the seed slurry must be agitated beforehand with a power of 10 $kW.m^{-3}$.

3.4.4. *Dewatering the massecuites*

The two main dewatering parameters are the liquid-phase viscosity and the non-sugar volume of the crystal products.

An 8°C preheating can reduce viscosity by 50%. We can also dilute the massecuite with a juice of limited viscosity. These steps are useful to ensure that the thick juice contains 62% of dry matter. In theory, the liquid-phase viscosity must not exceed 5–10 Pa.s.

Non-sugars are present in crystals in two ways:

– in mother liquor inclusions;

– on the crystal surface (adsorption).

Non-sugars can be eliminated from the crystal surface by preheating, which leads to their dissolution. We can also eliminate them by washing, that is, pulverizing water onto the crystal within the dewatering process.

The presence of agglomerates (due to overly fast crystallization, that is, to excessive supersaturation) is problematic, as they surround the mother liquor with impurities, and agglomerates are more difficult to wash than isolated crystals. Accordingly, crystal purity depends on the quality of crystallization. We also should not forget that heavy compounds (responsible for coloration) are the most difficult to eliminate, as they are poorly diffused in the washing liquor.

Like agglomerates, inclusions benefit from an overly fast crystallization (excessively high supersaturation). Inclusions are not eliminated by dewatering.

The presence of fine sugar ($d_p < 1$ μm) that passes through the continuous dewatering grid can explain the atypically high purity of molasses derived from C magma. Indeed, continuous dewatering centrifuges are equipped with grids rather than cloth, and can provoke splintering and breakage of fragile crystals with the production of fine sugar.

3.4.5. *Sugar refining*

Raw sugar contains between 1 and 3% of non-sugar that must be eliminated.

Filtering is the first step. This consists of eliminating the non-sugar present in the liquid surrounding the crystals. The sugar is mixed with syrup and dewatered. In this manner, we obtain "refined" sugar.

The sugar is then dissolved and purified by:

– carbonation or phosphatation;

– discoloration with charcoal or ion exchange.

The syrup obtained, the "fine liquor" is subjected to multilevel crystallization by water vaporization. We then obtain "refined white sugar".

As a general rule, beetroot sugar purity is more readily obtained than that of cane sugar.

3.4.6. *Thermal exchange data*

Owing to the high viscosity at 40°C, the cooling process is performed by circulating magma in tubes of diameter 10–15 cm at a velocity above 0.1 $m.s^{-1}$. The tubes are arranged in a triangle and the global transfer coefficient varies between 20 and 50 $W.m^{-2}.K^{-1}$. Water circulates on the outside of these tubes and, where appropriate air is used, with finned tubes instead.

On the other hand, the vaporization of water at 70°C occurs in the crystallizers with an internal exchanger, with short tubes of interior diameter 1–1.5 cm and length 1–1.5 m . Circulation requires a draught tube fitted with a marine impeller, with the magma circulating from the top to the bottom.

4

Crystallizers: Design and Dimensions

4.1. Introduction

4.1.1. *Qualities of a crystallized product*

These qualities concern size, granulometry, purity and facies.

1) Size is the result of a compromise, since regular crystals of large size are expensive to produce. Typically, we accept a size that allows easy dewatering (or filtration), storage and transport. Occasionally, a small size is preferable (pigments and charges for plastic materials, later dissolution).

2) The size distribution required is often as tight as possible, though this is not always the case.

3) Purity is attained by the careful choice of bleeding.

4) Facies (shape) is influenced by both the impurities present and supersaturation (dendrite growth for high supersaturations).

4.1.2. *Crystallizers*

We will not attempt to apply a classification to crystallizers, which would be a futile exercise. However, we will review the main types of devices currently in use, with a particular focus on their operating mechanisms.

This study will lead us to establish several simple criteria for choosing the crystallizer best adapted to the work in question.

4.1.3. *Overall balance of the crystallizer population*

Band $(L, L + \Delta L)$ sees its population varying during time δt.

Crystals of length L grow from $G(L)\delta t$ and enter the number section of $n(L)\, G(L)\, \delta t = nG\delta t$.

$n(L)\Delta L$ is the number of crystals present in 1 m^3 of slurry, between length L and $L + \Delta L$.

Crystals of length $L + \Delta L$ also grow, leaving their section. Their growth is:

$$G(L+\Delta L)\delta t$$

They leave their section at number:

$$n(L+\Delta L)G(L+\Delta L)\delta t = (n+\delta n)(G+\delta G)\delta t$$

Therefore, disregarding infinitely small terms of the second order:

$$(n\,\delta G + G\,\delta n + nG)\delta t$$

The overall balance of the band is (entry minus exit):

$$-\delta(nG)\delta t$$

Assuming a homogenous crystallizer, the population variation of the preceding band over time δt is:

$$V\,\delta n\,\Delta L$$

V is the crystallizer volume (m^3).

If there is vaporization, the variation is:

$$\delta V\, n\,\Delta L$$

Therefore, globally:

$$\Delta L\,\delta(nV)$$

The balance for the device is written:

$$\Delta L\,\delta(nV) + V\,\delta t\,\delta(nG) = 0$$

Hence:

$$\frac{\partial(nV)}{\partial t} + \frac{V\partial(nG)}{\partial L} = 0$$

However, this balance is incomplete.

We must account for:

1) the disappearance of D crystals per unit of time and unit of volume of the crystallizer. This disappearance is due to dissolution;

2) the apparition J of crystals per unit of time and unit of volume of the crystallizer. This apparition is due to primary and/or secondary nucleation (attrition);

3) the extraction of production q_s (L) n_s(L). The extracted flow varies with the size of crystals as their residence times are not necessarily identical;

4) the arrival of supply $q_a n_a$(L). However, most of the time, supply does not include crystals unless there is inflow of recycling returns;

The complete balance is written as:

$$\frac{\partial(nV)}{\partial t} + \frac{V\partial(nG)}{\partial L} + V\left[D(L) - J(L)\right] + q_s(L)n_s(L) = q_a n_a(L)$$

The dimensions in this balance are as follows:

$$\frac{\text{particles}}{\text{time} \times \text{length}}$$

4.1.4. *Balance of population with attrition*

The frequency of impacts between the impeller and the parent crystals is proportional to the number of these crystals coming into contact with the impeller:

$$f \propto Qc$$

Q: recirculation flow in the vat ($m^3.s^{-1}$)

c: number of particles per cubic meter of the suspension (m^{-3})

By applying probability p, so that an impact is followed by attrition:

$$f = pQc \qquad (0 < p < 1)$$

The apparition of fragments between L_a and $L_a + \Delta L_a$ in dimension is given by:

$$A = fm(L_a)\Delta L_a$$

$m(L_a)$: frequency of the number of fragments issued by a single impact (m^{-1})

The attrition parameters are ultimately:

– frequency of the number of fragments of size L:

$$m(L) = ML^{-3.25} \quad \text{with} \quad \int_0^\infty m(L)dL = 1$$

– growth decreases with the size of crystals, as attrition is equivalent to a negative growth proportional to the square of the crystal dimension:

$$G = G_o\left(1 - (L/L^*)^2\right)$$

A crystallizer's general balance of attrition for continuous crystallization is written as:

$$G\frac{\partial n}{\partial L} - G_o n 2L / L^{*2} + \frac{n}{\tau} = \frac{fM}{V_{sus}} L^{-3.25} = KL^{-3.25}$$

In addition, for batch crystallization:

$$\frac{\partial n}{\partial t}+G\frac{\partial n}{\partial L}=G_o n2L/L^{*2}+KL^{-3.25}$$

4.1.5. *Continuous homogenous crystallizer with attrition*

By bringing together the fragments resulting from attrition and crystals coming from primary nucleation or seeding into a single equation, we obtain:

$$G\frac{d(Ln(n))}{dL}=G_o\times\frac{2LM}{L^{*2}}+\frac{KL^{-3.25}}{n}-\frac{1}{\tau}$$

where L_{min} is the minimum size (that of seeds) and, if L nears L_{min}, the second term is dominant and strongly positive.

However, for low values of L, the derivative is cancelled where:

$$KL^{-3.25}=\frac{n}{\tau}$$

It then becomes negative.

L_{max} is the maximum size allowed and, if L nears this value, the derivative of Ln(n) is cancelled for:

$$\frac{1}{\tau}-\frac{G_o 2LM}{L^2_{max}}=0 \qquad \text{that is} \qquad L_{max}>2G_o\tau M$$

For example, if:

$$L^*=10^{-3}\,m \qquad G_0=10^{-7}\,m.s^{-1} \qquad \tau=4000\ s \qquad M=0.5$$

We should then have:

$$10^{-3}>2\times10^{-7}\times4.10^3\times0.5=4.10^{-4}$$

Finally, variations of Ln (n) will appear as in Figure 4.1.

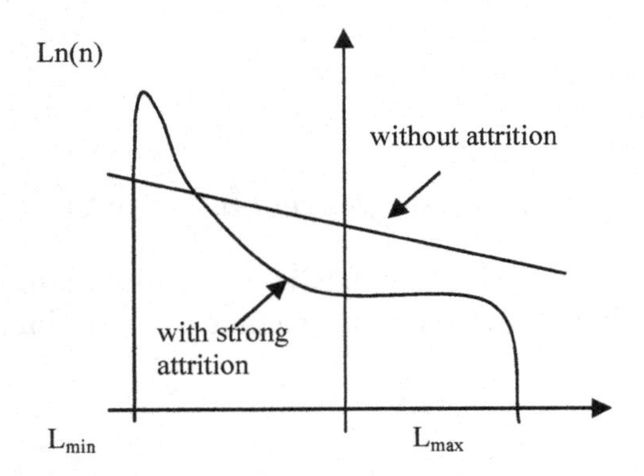

Figure 4.1. *Ln (n) variations with the crystal size*

4.1.6. *Batch with attrition*

The balance equation becomes:

$$\frac{\partial n}{\partial t}+G\frac{\partial n}{\partial L}=nG_o\times\frac{2L}{L_{max}^2}+KL^{-3.25}=f(n,L)$$

This type of equation is resolved by the characteristic method [NOU 85].

$$\frac{dt}{1}=\frac{dL}{G}=\frac{dn}{f}$$

Therefore:

$$L=L_o+Gt \qquad (L_o\neq 0 \text{ if the batch is seeded}).$$

Moreover:

$$\frac{dn}{dL}=\frac{f(n,L)}{G}=n\times\frac{2L}{L^{*2}\left(1-\frac{L^2}{L^{*2}}\right)}+\frac{KL^{-3.25}}{G_o\left[1-\left(L/L^*\right)^2\right]}$$

We observe that n is an increasing function of L. Remember that the total number of crystals is:

$$N = \int_{L_{min}}^{L_{max}} n(L)dL$$

NOTE.–

Irrespective of whether this n is for continuous crystallization or for batch crystallization, both of these variables are the solutions of a differential equation that is linear in form.

$$y' + y P(x) = Q(x)$$

For which the general solution is [SPI 92]:

$$y = \exp\left[\int P(x)dx\right]\int Q(x)\exp\left[\int P(x)dx\right]dx + cste$$

4.1.7. *Agitation*

In a crystallizer with a draft tube, we typically use marine impellers located on the lower part of the draft tube. The draft tube is the dispositive allowing the maximum flow for minimum energy. The drop in pressure through a draft tube is for the most part attributed to the turning at its ends.

$$\Delta P = k\frac{\rho_B V^2}{2} \quad (\rho_B \text{ : density of the slurry})$$

where V is the velocity in the tube.

For a simple tube, k is between 1.7 and 2.

By shaping the lower part of the tube, that is, the tube entrance (where the impeller is present), coefficient k can be brought down to 1.4 (according to Oldshue [OLD 83]).

The role of agitation is, of course, to keep the crystals in suspension, but above all to homogenize the slurry and avoid excessive local supersaturation.

The agitator must not break the crystals, otherwise fines will appear, which will affect the size distribution. To address this, experience shows that the velocity at the blade ends must be limited. Nonetheless, while excessive velocity can damage crystals, vigorous agitation can round their aspect, making them purer and more robust.

In general, baffles are significant in the agitation process, while in crystallization, even if they are several centimeters from the wall, they can lead to the accumulation of crystals, provoking pipe blockages. This is one of the advantages of a draft tube system that avoids baffles.

In case of external recirculation, the circulation flow can reach several hundred or even several thousand cubic meters per hour.

The pumps used are axial with low manometric height and do not cause too much crystal damage.

4.1.8. *Thermal exchange*

If we require a crystallizer that is significantly greater in size than the pilot device, while keeping the same thermal program, we are obliged to introduce a heating coil into the vat.

Unfortunately, in most cases, a crystal deposit develops on the outside of the heating coil despite agitation. This is the phenomenon of crusting, which is detrimental to thermal transfer. The most common solution consists of recirculating the slurry through the external exchanger using a pump. The exchanger is used if the solubility varies significantly with temperature, and also acts as a means of heating the solution to evaporate the solvent should solubility vary little with temperature.

It could be asked why it is not more worthwhile to recirculate clear liquor. In fact, this can be done in case of evaporation, but for cooling, if we wish to avoid creating fines, supersaturation without crystals has to last for the time for the liquor to join the useful volume of the crystallizer. This time must not exceed the latency. Indeed, it increases with installation size. For these reasons, we simply circulate slurry in the external exchanger. In this way, the solid can be deposited immediately on the crystals rather than on the surfaces of thermal transfer, or rather than multiplying the seeds.

The exchanger is of the tubular type with one or two passes according to the flow. We must have easy access to clean the tubes in case of crusting.

With this type of exchanger, several hundred square meters can be reached. Blockages can be averted by selecting tubes of sufficient diameter (40 mm). On the other hand, the slurry velocity must be in the order of 1.5–2.2 m/s. The thermal transfer calculation for slurry is performed by adopting the physical properties of the liquid phase. Indeed, the presence of crystals scrapes and activates the boundary layer. In addition, the thermal conductivity of compact solids is clearly greater than that of liquids. Accordingly, this method favors reliability.

When the exchanger is used for heating, we must avoid boiling in the tubes. The temperature of the wall in contact with the slurry corresponds to a certain vapor pressure that must be lower than the hydrostatic pressure governing the upper part of the exchanger.

During crystallization by cooling, tube crusting can occur if the difference in temperature between the wall surface and the slurry exceeds a critical value ΔT_{cr} according to:

– the nature of the solvent–solute and the width of the metastable zone in particular;

– slurry temperature;

– velocity in the tubes (ΔT_{cr} increases with this velocity);

– the nature of the wall (roughness). A rough surface benefits crusting.

One solution consists of circulating the cooling fluid in an intermediate loop to ensure that the temperature difference between the slurry and the intermediate loop remains below 3 or 4°C.

4.2. Continuous homogeneous crystallizer (CHC)

4.2.1. *CHC definition*

The device is perfectly mixed. Throughout the useful volume (occupied by crystal slurry), the liquor composition is the same and the crystal concentration is also the same throughout; the device is equipped with:

– a supply;

– a slurry exit;

– (possibly) a water vapor exit if we proceed by vaporization.

4.2.2. *Crystallizer with a draft tube and an internal exchanger*

The industrial system that leads to the lowest attrition is the draft tube crystallizer, with an internal exchanger as represented in Figure 4.2. This design is applicable to both vaporization and cooling.

The draft tube extends above the body, as a significant level of slurry above the body is required in case of vaporization in order to avoid boiling in the tubes. The diameter of the draft tube is a third or half that of the crystallizer itself.

For a given velocity in the tubes (1.5 $m.s^{-1}$), the thermal transfer and surface transfer coefficients are invariable. We can then choose between:

– many short tubes (length: 1.5 m) and consequently a significant circulating flow;

– few long tubes (length: 4–7 m) and a lower circulating flow.

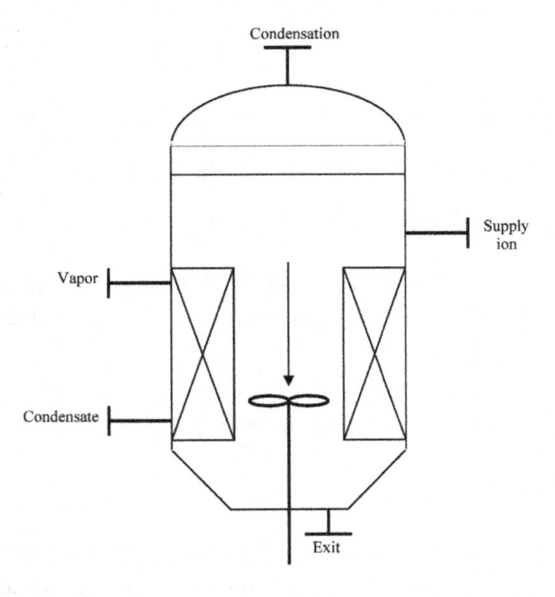

Figure 4.2. *A crystallizer with a draft tube and an internal exchanger*

The first solution is appropriate for the CHC, and is also more expensive (due to the price of tube plates). On the other hand, the discharge pressure required for the impellor is lower, and attrition less significant. The second solution makes the crystallizer closer to an external exchanger and is also less expensive. In reality, the shorter the tubes (due to the drop in pressure), the higher the magma's viscosity (which is high in the case of sugar)

We should note here that crystallizers with an internal heat exchanger are more of an economical investment than devices with an external exchanger (that is, with forced circulation).

4.2.3. *Definition of draft tube*

The tube diameter is 30–50% that of the vaporization space. The distance from the bottom of the tube to the bottom of the vat is equal to the diameter D_T of the draft tube. In order to alleviate the vortex effect, the distance from the top of the tube to the free surface of the liquid considered as at rest is equal to $D_T/2$. In the work of Oldshue (p. 479, [OLD 83]), we find the plan of a draft tube. To reduce the drop in pressure, it is advisable to "round" the lower part of the tube by fitting it with an external ring made with a tube of diameter $D_T/4$.

The draft tube fitted with a marine impeller is capable of a significant recirculation flow. This flow must meet two conditions:

– avoiding dividing the crystals, that is, ensuring that their retention time does not vary according to size:

$$Q \geq 0.01 V_c$$

V_c : useful volume of the crystallizer (m^3)

Q : recirculated flow ($m^3.s^{-1}$);

– avoiding supersaturation heterogeneities, that is, ensuring that the composition of the liquor is the same throughout the crystallization body:

$$Q \geq \frac{W}{C_B \rho_B \Delta T}$$

ΔT: fictive variation of temperature between 0.5 and 1°C

ρ_B: slurry density (kg.m^{-3})

C_B: thermal capacity of the slurry (J.kg^{-1}.°C^{-1})

W: thermal power injected into the crystallizer (W).

The clean liquor is extracted from the upper part of the clarification zone, which avoids attrition of the crystals by the pump. The liquor leaving the exchanger is injected at the base of the crystallizer, that is, near the large crystals (maximum crystal surface) that grow by absorbing the supersaturation, which avoids excessive nucleation.

A reasonable value for slurry velocity in the draft tube is 2 m/s. The liquid rises in the tube.

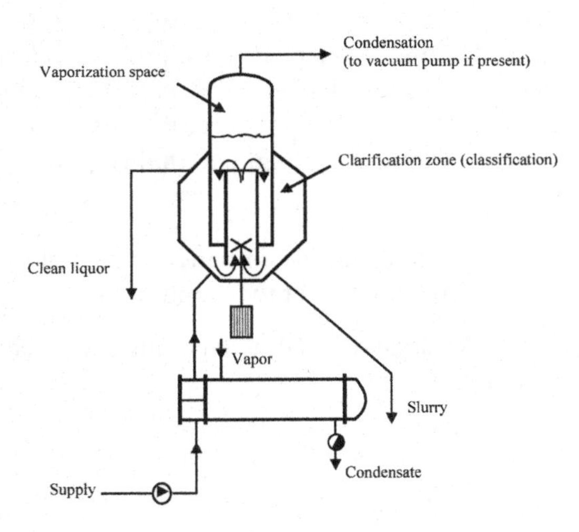

Figure 4.3. *A crystallizer with a draft tube and vaporization (Swenson)*

The impeller, of the marine type with three vanes, is placed on the lower part of the tube (at a third of its height).

According to the classic formula, the flow created by an impeller at a pitch equal to p is:

$$Q = 0.5 \times A \times N \times p$$

A: cross-section of the draft tube (m^2)

N: rotation speed ($rev.s^{-1}$)

p: impeller pitch (often equal to its diameter) (m)

This equation provides the impeller rotation speed.

To ensure correct homogeneity and avoid unscheduled decantation, the speed at the end of each vane must be greater than 3 $m.s^{-1}$. This speed must not exceed 7 $m.s^{-1}$ for brittle crystals and can reach 10 $m.s^{-1}$ for resistant particles. In order to diminish the shear at the end of the vanes, we leave a clearance of 2–3 cm between the impeller and the tube wall. The impeller's position in the tube is not significant; however, if we wish to correct an unscheduled decantation that occurs during a stoppage, the impeller should preferably be placed on the lower part as mentioned earlier in this chapter.

According to Oldshue's formula [OLD 83], the pressure loss created by circulation is:

$$\Delta P = 1.5\frac{\rho_B V^2}{2}$$

ΔP: pressure drop (Pa)

ρ_B: density of slurry ($kg.m^{-3}$)

V: velocity in tube ($m.s^{-1}$).

The energetic yield η of a marine impeller operating in these conditions is approximately 20%, so that the shaft power is:

$$P_a = \frac{Q\Delta P}{0.2}$$

P_a: shaft power (W).

The power number N_P can then be assessed. Indeed:

$$P_a = \frac{Q\Delta P}{\eta} = \frac{0.5}{0.20}\frac{\pi}{4}d^3N\left(\frac{p}{d}\right)\times 1.5\rho_B \times (0.5)^2 \times (Nd)^2\left(\frac{p}{d}\right)^2$$

Therefore:

$$P_a = 0,77 \ N^3 d^5 (p/d)^3$$

That is:

$$N_p = 0.77$$

This value is greater than the value of 0.43 for an impeller operating without a draft tube. However, the tube has the advantage of orientating the circulation vertically, which effectively neutralizes the decantation effect due to gravity.

If vaporization occurs, the top of the draft tube must be 30 cm below the liquid level to avoid splashes and contamination on the wall of the vaporization space.

In some cases, magma can be particularly viscous. Here, its viscosity is given by Bruhns' formula [BRU 96]:

$$\frac{\mu_M}{\mu_{LM}} = 1 + 2.8\left(\frac{\varphi_T}{\varphi_{max} - \varphi_T}\right)^{4/3}$$

μ_M and μ_{LM}: viscosities of magma and the mother liquor (Pa.s)

φ_T: volume fraction of solid in the magma

φ_{max}: maximum value of φ_T (complement to 1 of resting crystals porosity)

For near-equidimensional crystals (with their three main dimensions of the same order), we take $\varphi_{max} = 0.6$

This expression is not applicable for particles below 50 nm.

For plates, we take φ_{max} # 0.20

For fibers, we take $0.01 < \varphi_{max} < 0.05$

The above is only applicable for magmas that are not too viscous, such as those present in most crystallizers. In such cases, we use the pressure drop formulae for pipes both for the draft tube and for the thermal exchanger. Therefore, we have:

$$\varphi_T \leq 0.2$$

$$\mu_{LM} \leq 0.1 \text{ Pa.s}$$

This is not the case in sugar refineries where, at 40°C, and for C sugar (after useful production), we can have:

$$\varphi_T = 0.56$$

$$\mu_{LM} = 1 \text{ Pa.s}$$

The formula above gives 96 Pa.s, while the viscosity of magma can reach 1,000 Pa.s (though this is simply an apparent viscosity as magma is no longer Newtonian).

4.2.4. *Classification zone*

We will provide the elements in order to simulate the operation of a classification zone.

1) Sedimentation rate. We will use the following values:

ε: volume fraction of the liquid

$v\ell$: maximum drop velocity of an isolated particle ($m.s^{-1}$)

$$v_\ell = \frac{g d^2 \Delta\rho}{18\,\mu}$$

d: minimum dimension of a crystal

μ: viscosity of the liquid phase (Pa.s)

$\Delta\rho$: difference in solid and liquid densities ($kg.m^{-3}$)

g: acceleration due to gravity ($9.81\ m.s^{-2}$)

The sedimentation rate is:

$$v = v_1 \varepsilon^n$$

We calculate:

$$Re = \frac{v_1 d \rho_L}{\mu}$$

$0.2 < Re < 1 \quad n = 4.4 / Re^{0,023}$

$1 < Re < 500 \quad n = 4.4 / Re^{0,0982}$

$Re > 500 \qquad n = 2.39$

The rate calculated this way only applies to spherical particles. For angular particles such as crystals, this rate must be divided by 2.

2) Rate and residence time within a section. We divide the classification zone into *n sections of current index j rising from 1 to n from the bottom to the top of the classification zone*. The bands' widths are all equal to Δz. If v_0 is the rate in an empty vat measured upwards and positively, and if v_{ij} is the sedimentation rate of class i particles of diameter d_{ij} in section j, the absolute rate of the particle relative to the workshop is:

$$v_{aij} = v_0 - v_{ij}$$

With:

$v_{aij} > 0$, the particle enters section j + 1

$v_{aij} < 0$, the particle enters section j – 1

The retention time within a section is: $\Delta\tau_{ij} = \Delta z / v_{aij}$

3) Growth of crystals within section j. If d_{ije} is the size of particles entering section j (and coming from sections j + 1 and j – 1), the crystal size exiting the section is:

$$d_{ijs} = d_{ije} + G\,\Delta\tau_{ij}$$

In sections j – 1 and j + 1 into which the particles from index i would have entered, we can calculate: $V_{a, i, j+1}$ or $V_{a, i, j-1}$

4) The main purpose of the calculation is knowing the cutoff diameter of the classification zone, that is, the size of particle d_{io}^* entering at the base so that, after growth, at the summit we have $z = n\Delta z$.

$$v_{ai} = 0$$

To this end, we have a liquid fraction ε that is constant and equal to the value of supply at the base of the zone. We also have constant supersaturation and growth.

The cancellation of absolute rate v_{ain} gives a cutoff diameter of d_{in}^*. Going back in time, we recognize the value d_{io}^* from the diameter of the same particle at the entry of the classification zone.

If $d_{io} < d_{io}^*$ the particle is carried with the "clean" liquor.

If $d_{io} > d_{io}^*$ the particle remains in the crystallizer body.

4.2.5. *Purpose of classification*

The purpose of classification is twofold:

– to hinder the apparition of fines;

– to eliminate existing fines.

The separate decanting of clarified liquor automatically leads to an increase in the crystal content of the magma. The crystal surface available for growth is increased and will absorb the liquor's supersaturation to the detriment of both nucleation and the apparition of fines. The solid-phase content should preferably reach 20% in volume but should not exceed this value in order to avoid too much attrition in the draft tube.

The decanting of clarified liquor also includes the undesirable fines, which can then be:

– dissolved by heating the liquor;

– separated in a centrifugal screw decanter with horizontal axis. The sludge obtained is reinjected into the supply. However, this solution is expensive in terms of investment.

4.2.6. *Theory of CHC without attrition*

If we assume that the residence time is identical for both the crystals and the solution, this implies that the slurry decanting must be performed in an isokinetic manner (the crystals move at the same velocity as the liquor). Therefore, the balance equation is:

$$\frac{\partial(Gn)}{\partial L}+\frac{n}{\tau}=0$$

For dimensions greater than 100 nm and below 2 mm, growth often depends less on the crystals' size. Therefore, the balance equation is easily integrated and we obtain:

$$\mathrm{Ln}(n)=\mathrm{Ln}(n_o)-\frac{L}{G\tau}$$

This result can be obtained in the laboratory in a crystallizer of several liters. However, according to Mersmann [MER 01], some precautions are indispensable in order to avoid attrition:

0.5 W/kg of suspension is the upper limit of the agitation power density = $\overline{\varepsilon}$.

"density" (mass concentration)	$m_T < 50\,kg.m^{-3}$
Volume fraction of crystals:	$\varphi_T < 0.02$
Residence time	$< 5{,}000$ s

The origin ordinate e and the slope of the straight line Ln(n) = f(L) provide n_o and G. We may deduce:

$$N_T = \int_0^{\infty} n_o \exp\left(-\frac{L}{G\tau}\right) dL = n_o G\tau$$ (number of crystals per cubic meter of suspension)

The nucleation rate J per unit of time and per cubic meter of suspension is:

$$J = \frac{dN}{dt} = \left(\frac{dN}{dL}\right)_{L=0} \times \left(\frac{dL}{dt}\right)_{L=0} = n_o G$$

Nyvlt attempted to measure J by direct observation of the apparition of the first seeds, but this particularly dubious method was abandoned.

Note that, in the absence of attrition, the expression of J according to the derivatives taken for L = 0 is universal, irrespective of the n variation law as a function of L.

4.2.7. *Other values*

1) Crystal surface per cubic meter of the slurry:

$$A(L) = \beta \int_0^L L^2 n \, dl$$

$$A(L) = 2\beta n_o (G\tau)^3 \left[1 - \left(\exp\left(\frac{-L}{G\tau}\right)\right)\left(1 + \frac{L}{G\tau} + \frac{1}{2}\left(\frac{L}{G\tau}\right)^2\right)\right]$$

The volume area is obtained by integrating from zero to infinity:

$$A_c = \beta\mu_2 = 2\beta \, n_o (G\tau)^3 \qquad (m^2.m^{-3})$$

2) Crystal mass per cubic meter of slurry (slurry "density"):

$$M(L) = \alpha\rho_c \int_0^L L^3 n \, dl$$

$$M(L) = 6\alpha\rho_c n_o (G\tau)^4 \left[1 - \left(\exp\left(\frac{-L}{G\tau}\right)\right)\left(1 + \frac{L}{G\tau} + \frac{1}{2}\left(\frac{L}{G\tau}\right)^2 + \frac{1}{6}\left(\frac{L}{G\tau}\right)\right)\right]$$

The total mass of crystals per cubic meter of slurry is:

$$M_c = 6\alpha\rho_c n_o (G\tau)^4 \qquad (\text{kg.m}^{-3})$$

3) Dominant size of crystals (distribution mode):

The distribution density as mass is:

$$\frac{dM}{dL} = \alpha\rho_c L^3 n$$

Applied to total mass, we obtain:

$$w(L) = \frac{1}{M_c}\frac{dM}{dL} = \frac{L^3 n}{6(G\tau)^4} = \frac{L^3 n_o \exp\left(\frac{-L}{G\tau}\right)}{6(G\tau)^4}$$

The dominant size maximizing w (L) is:

$$6(G\tau)^4 \frac{dw(L)}{dL} = \left(3L^2 - \frac{L^3}{G\tau}\right) n_o \exp\left(\frac{-L}{G\tau}\right) = 0$$

Hence: $L_D = 3G\tau$

Note that the crystal surface A_c can be expressed by:

$$A_c = \frac{M_c\beta}{\alpha\rho L_D}$$

4) Mean crystal size:

$$\bar{L} = \frac{\mu_4}{\mu_3} = \frac{\int_0^\infty L^4 n\, dL}{\int_0^\infty L^3 n\, dL} = 4G\tau$$

5) Mean crystal mass:

$$\bar{m} = \frac{M_c}{N} = 6\alpha\rho_c (G\tau)^3 = \frac{2}{9}\alpha\rho L_D^3$$

6) Median dimension:

The median is defined by:

$$0,5 = \frac{\int_0^{L_{50}} L^3 n \, dL}{\int_0^{\infty} L^3 n \, dL}$$

hence:

$$L_{50} = 3.67 G\tau$$

7) Variation coefficient:

It is defined by:

$$C.V. = 100 \frac{\left[\frac{\mu_5}{\mu_3} - \left(\frac{\mu_4}{\mu_3}\right)^2\right]^{1/2}}{\frac{\mu_4}{\mu_3}}$$

For the CHC:

$$C.V = 50\%$$

8) Volume of crystals per cubic meter of suspension:

$$\varphi_T = \frac{M_c}{\rho_c} = 6\alpha n_o (G\tau)^4$$

NOTE.–

The value M_c is incorrectly known as the suspension "density". In fact, the suspension density is:

$$\rho_{sus} = \varphi_T \rho_c + (1 - \varphi_T)\rho_L$$

ρ_c and ρ_L : density of crystals and liquor ($kg.m^{-3}$)

EXAMPLE.–

Here, we will provide values of the characteristic parameters of crystallization in order to familiarize the reader with the orders of magnitude.

1) Nucleation rate:

$$J = 5.52.10^6\, M_c\, \varepsilon\, \Delta X^5$$

Magnitude X is the ratio of solute in the liquor:

$$X = \frac{\text{solute mass}}{\text{solvent mass}}$$

$$\Delta X = X - X^*$$

with:

$$M_c = 466 \text{ kg} / \text{m}^3 \qquad \bar{\varepsilon} = 500 \text{ W.m}^{-3} \qquad \Delta X = 0.06$$

X^* is the level of solute on saturation.

We find:

$$J = 10^6 \text{ seeds.m}^{-3}\text{.s}^{-1}$$

2) Growth rate:

$$\frac{3\alpha\rho_s G}{\beta} = k_G \Delta X^g$$

with:

$\alpha = 0.42$ $\beta = 3.4$ $\rho_s = 1,700 \text{ kg/m}^3$

$k_G = 8.57.10^{-3} \text{kg/m}^2.\text{s}$ $\Delta X = 0.06$ $g = 1.5$

we find:

$$G = 2.10^{-8} \text{m.s}^{-1}$$

3) Population density for L = 0:

$$n_o = J/G = 10^6 / 2.10^{-8} = 5.10^{13} \text{seeds.m}^{-4}$$

4) Crystal concentration:

$$M_c = 6\,\alpha\,\rho_s\,n_0 (G\tau)^4$$

with the previous values and $\tau = 10,800$ s, we find: $M_c = 466 \text{ kg.m}^{-3}$

5) Volume area of crystals:

$$A_c = 2\beta\, n_0 (G\tau)^3$$

We find: $A_c = 3,426 \text{ m}^2.\text{m}^{-3}$

6) Dominant size of crystals:

$$\bar{L}_D = 3\,G\tau = 648.10^{-6} \text{m}$$

7) Slurry "density":

$D_B = 0.33$ $\left(\text{with } \rho_L = 1,300 \text{ kg.m}^{-3}\right)$

4.2.8. *Parameters to manage*

Let us remove τ and n_o between J, φ_T and L_{50}. We obtain:

$$L_{50} = 3.67\left(\frac{G\varphi_T}{6\alpha J}\right)^{0,25}$$

The crystallizer design allows for φ_T to be controlled together with supersaturation, that is, J and G, in order to obtain the median dimension required. However, in reality, things are not this simple. If we increase the residence time, supersaturation decreases, and consequently, the value of G is such that the median size does not increase in proportion to τ even when φ_T is proportional to τ^4. In fact, L_{50} slowly increases with τ and even begins to decrease if attrition becomes significant.

We observe that L_{50} varies similarly to φ_T at power 0.25. Accordingly, it is worthwhile for the crystal volume fraction φ_T to approach 0.20, that is, to approach its maximum value, for an acceptable agitation.

4.2.9. *CHC series without attrition*

For a single device, we know that the population density expression is:

$$n_1 = n_o \exp\left(-\frac{L}{G\tau}\right)$$

Let us now distinguish two cases:

1) Nucleation occurs only in the first crystallizer and the speed of growth is the same in all crystallizers.

In the second crystallizer, we can write out the following balance:

$$Q(n_2 - n_1)\Delta L = -V_c \Delta n_2 G$$

where V_c is the volume of a device so that:

$$\frac{dn_2}{dL} = \frac{1}{G\tau}(n_2 - n_1)$$

We can verify that the solution to this equation is:

$$n_2 = n_o \left(\frac{L}{G\tau}\right) \exp\left(\frac{-L}{G\tau}\right)$$

For k crystallizers, this can also be generalized like so:

$$n_k = n_o \left(\frac{L}{G\tau}\right)^{k-1} \frac{\exp\left(\frac{-L}{G\tau}\right)}{(k-1)!}$$

2) Nucleation and growth are identical in m crystallizers, so:

$$n_m = n_o \left[1 + \frac{L}{G\tau} + \frac{1}{2!}\left(\frac{L}{G\tau}\right)^2 + \dots\dots\dots + \frac{1}{(m-1)!}\left(\frac{L}{G\tau}\right)^{m-1}\right] \exp\left(\frac{-L}{G\tau}\right)$$

This expression can be deduced from the previous case by noting that population is the total of m populations corresponding to k, varying by 1 to m.

Randolph and Larson [RAN 71] show that, in the two previous cases, the size of crystals is lower and distribution is tighter compared with the case of a single crystallizer. These authors give the surface distributions and the mass for two crystallizers.

Often, it is not for reasons of size distribution that crystallizers are placed in series, but to save energy:

– evaporation for several effects;

– cooling on two levels. The cooler levels operate with a refrigeration unit and the entry level is cooled by the mother liquor from the cooler level.

However, the abundance of slurry pumps can create attrition problems for certain products.

4.2.10. *Operative relationship of a CHC without attrition*

Irrespective of the crystal size, the rate of increase in mass is:

$$\frac{dm}{dt} = k_G \beta L^2 \Delta X^g = \frac{d}{dt}\left(\alpha \rho L^3\right) = 3\alpha \rho G L^2$$

Therefore:

$$\frac{k_G \beta \Delta X^g}{3\alpha \rho G} = 1$$

However (see section 4.2.7):

$$\frac{A_c}{M_c} = \frac{3\beta n_o (G\tau)^3}{6\alpha \rho n_o (G\tau)^4} = \frac{\beta}{3\alpha \rho G\tau} = \frac{\beta}{\alpha \rho L_D}$$

The crystallizer's productivity is:

$$P = k_G A_c \Delta X^g = \frac{k_G M_c \beta \Delta X^g}{\alpha \rho L_D}$$

The nucleation rate is equal to the productivity quotient (in kg.s^{-1}) by the mean crystal mass (see section 4.2.7).

$$J = \frac{9P}{2\alpha \rho L_D^3}$$

However, we frequently write:

$$J = k_N^* M_c^j \varepsilon^h \Delta X^n$$

$\overline{\varepsilon}$: agitation power per cubic meter of slurry (W.m^{-3}).

By equating the two expressions of J, we obtain:

$$\Delta X = \left[\frac{9P}{2\alpha \rho L_D^3 k_N^* M_c^j \varepsilon^h}\right]^{1/n}$$

Let us replace ΔX with its value in the expression of P:

$$P = \frac{k_G M_c \beta}{\alpha \rho L_D} \left[\frac{9P}{2 \alpha \rho L_D^3 k_N^* M_c^j \Sigma^h} \right]^{-g/n}$$

Or again:

$$L_D^{\left(1+\frac{3g}{n}\right)} = \frac{k_G \beta}{\alpha \rho} \left[\frac{9}{2 \alpha \rho k_N^* \varepsilon^h} \right]^{g/n} \times M_c^{\left(1-\frac{jg}{n}\right)} \times P^{\left(\frac{g}{n}-1\right)}$$

Lifting the two members of the previous equation to power $i = \dfrac{n}{g}$ comes to:

$$L_D^{(i+3)} = \frac{9}{2 \alpha \rho k_N^* \varepsilon^h} \left[\frac{k_G \beta}{\alpha \rho} \right]^i \times M_c^{i-j} \times P^{1-i} = B \times \frac{M_c^{i-j}}{p^{i-1}} \qquad [4.1bis]$$

However, the M_c / P ratio is written:

$$\frac{M_c}{P} = \frac{\alpha \rho L_D}{k_G \beta \Delta X^g} = \frac{3 \alpha \rho G \tau}{k_G \beta \Delta X^g} = \tau \qquad [4.2]$$

By eliminating P between [4.1bis] and [4.2], we obtain the desired relationship:

$$L_D = B^{\frac{1}{L+3}} \times M_c^{\frac{i-j}{i+2}} \times \tau^{\frac{i-1}{i+3}} \qquad [4.3]$$

In general, j is close to i, and M_c scarcely intervenes in the equation [4.3].

Often, the value of i, greater than 1, is of the order of 2 and we have:

$$L_D \propto \tau^{0.2}$$

We observe that the crystal size is far from proportional to the residence time. This can be explained by the supersaturation ΔX decreasing with residence time.

4.2.11. *Installation possibilities of a CHC (population density)*

1) Separate decanting of clean liquor.

Decanting the product at flow Q_P is performed on the lower part of the *continuous homogenous* crystallizer. Decanting of clean liquor is performed separately at flow Q_{CL} at the upper part of the classification zone.

The cutoff dimension is L_F between the fines (decanted at flow Q_P + Q_{CL}) and the product decanted at Q_P. The corresponding residence times are:

$$L < L_F \qquad \frac{1}{\tau_F} = \frac{Q_P + Q_{CL}}{V}$$

with $$\frac{\tau_P}{\tau_F} = \frac{Q_P + Q_{CL}}{Q_P} = R = \frac{Q_T}{Q_P} = \frac{\tau_P}{\tau_T}$$

$$L > L_F \qquad \frac{1}{\tau_P} = \frac{Q_P}{V}$$

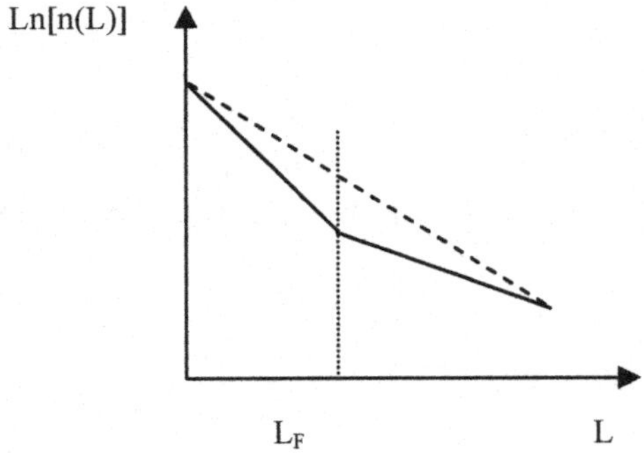

Ln[n(L)]

L_F L

Figure 4.4. *Separate decanting of clean liquor*

As the dominant crystal size of the product is equal to $L_D = 3G\tau$, we can expect that this size will be increased as:

$$\tau_P = R\tau_T$$

However, as growth G decreases little with residence time, the crystal size will increase less than proportionally with τ.

2) Destruction of fines:

V_{CLA} : overflow (fines + liquor)

Q_P: underflow (crystal slurry)

$$R = \frac{\dot{V}_P + \dot{V}_{CLA}}{\dot{V}_P} > 1$$

In the laboratory, with R = 2, we observe an increase of 15% in $L_{Dominant}$. The higher the R, the longer the residence time τ_D of the product. The residence time for fines decantation ($L < L_F$) is:

$$\tau_F = \frac{V}{\dot{V}_P + \dot{V}_O} = \frac{V}{R\dot{V}_P}$$

For the product ($L > L_F$):

$$\tau_P = \frac{V}{\dot{V}_P}$$

We accept that the n_o of seeds is identical with and without fines destruction. We also accept that τ_P is the same with and without fines destruction. However, growth G_F with fines destruction is higher than that of G_O without destruction as the crystals are larger.

Depending on whether L is greater than or less than L_F, the variation law of n according to L is different:

$$L < L_F \qquad n = n_o \exp\left(-\frac{RL}{G_F \tau_P}\right)$$

For $L = L_F$, there is no discontinuity of n(L). Accordingly:

$$L > L_F \qquad n = n_o \exp\left(-\frac{RL_F}{G_F \tau}\right) \exp\left(-\frac{L - L_F}{G_F \tau_P}\right)$$

G_F is greater than G_o which means that with fines destruction, supersaturation is increased.

Figure 4.5 represents what has been explained above.

$$\dot{V}_0 = (R - 1)\dot{V}$$

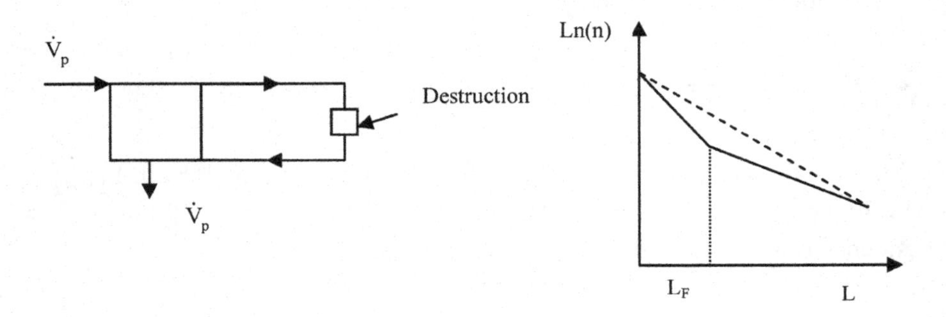

Figure 4.5. *Fines destruction*

Typically, the previous results are only verified for a continuous homogenous crystallizer without attrition, that is, in the laboratory.

3) Product classification:

After crystallizer decantation, slurry containing the product is cleared of fines in a classifying device that may be:

– an elutriation column;

– a hydrocyclone.

Compared with the crystallizer body, the product exits at the global flow zQ_P.

The fines leave in flow zV_P but return in flow $(z - 1)\ Q_P$, so, in total, decantation has the flow:

$$z\dot{V}_P - (z - 1)\dot{V}_P = \dot{V}_P$$

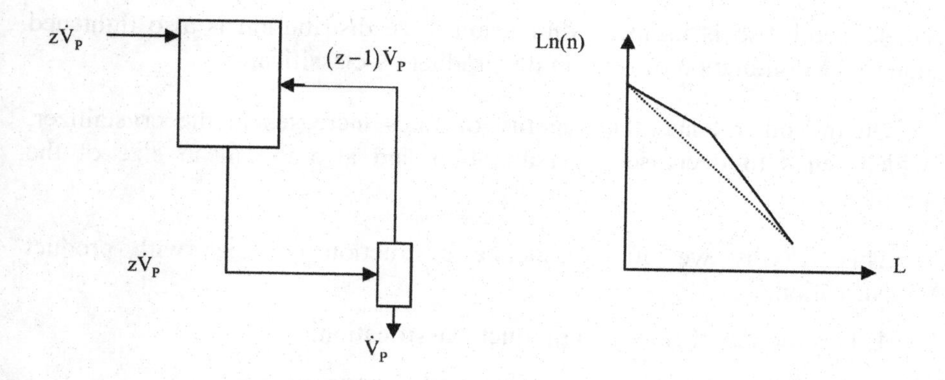

Figure 4.6. *Product classification*

The residence times are:

Fines $(L < L_C)$ $\tau_C = \frac{V}{\dot{V}_P}$

Product $(L < L_C)$ $\tau_C = \frac{V}{\dot{V}_P}$

L_C is the "cutoff dimension" of crystals in the classifier.

Hence, the expression of n(L):

Fines $L \leq L_C$ $n = n_o \exp\left(-\frac{L}{G\tau_C}\right)$

Product $L \geq L_C$ $n = n_o \exp\left(-\frac{L_C}{G\tau_C}\right)\exp\left(-\frac{z(L - L_C)}{G\tau_C}\right)$

Or rather: $n = n_o \exp\left(\frac{(z-1)L_C}{G\tau_C}\right)\exp\left(-\frac{zL}{G\tau_C}\right)$

Thus, the slope of the straight line corresponding to the product (Ln(n) = f(L)) is increased. The residence time of the product is reduced and consequently, so is its mean dimension. Size distribution is also tightened due to the elimination of fines in the product by classification.

On the other hand, the quantity of fines increases in the crystallizer, which tends to decrease supersaturation and also the mean size of the product.

This is why we often associate destruction of fines with product classification.

4) Destruction of fines and product classification:

The corresponding expressions for n = f (L) are:

$$L < L_F \qquad n = n_O \exp\left(-\frac{RL}{G\tau}\right)$$

$$L_F < L < L_C \qquad n = n_O \exp\left(-(R-1)\frac{L_F}{G\tau}\right)\exp\left(-\frac{L}{G\tau}\right)$$

$$L > L_C \qquad n = n_O \exp\left(-(R-1)\frac{L_F}{G\tau} + (z-1)\frac{L_C}{G\tau}\right)\exp\left(-\frac{zL}{G\tau}\right)$$

4.2.12. *Modulation of crystal content*

The slurry density can be altered in various ways. Relative to its natural density, we can increase or decrease it.

To increase the density, we decant the crystallizer's slurry and treat it in a clarifier. The clear overflow is evacuated and the thick underflow is recycled to the crystallizer.

Instead of a clarifier, we can also use:

– an undisturbed space inside the crystallizer (clarification zone);

– a centrifugal screw decanter with horizontal axis;

– a gravity decanter;

– a battery of hydrocyclones in series.

In order to decrease the slurry density, we decant it, treating it in a centrifuge. The crystals are evacuated to drying and the mother liquor is recycled to the crystallizer, following a purge to eliminate impurities, or to eliminate the mother liquor from the material balance if there is no evaporation. The greater the flow crossing the centrifuge or the filter, the lower the slurry density.

Often, we need to increase the slurry density in order to increase the crystallizer productivity (production per unit of volume). For this result, we proceed (as explained above) by separate decantation of the clean liquor. Indeed, in Figure 4.4, we observe that the crystal size decreases. Thus, Bubnik *et al.* [BUB 84] showed that in the crystallization of sucrose, an increase in the crystal weight ratio in the magma from 20 to 60% provokes a reduction by half in the crystal growth rate.

4.3. Continuous forced circulation crystallizer (CFCC)

4.3.1. *Description*

In the external loop, the slurry crosses a tube exchanger that provides the heat or cold as required. This exchanger can be a single tube pass (horizontal or vertical) or double pass (horizontal).

The slurry exits the crystallizer body by a funnel bottom and is directed to the exchanger by an axial pump (generating attrition). It is reintroduced into the body by one of the two standardized devices that will be described below.

Recirculation flow must be such that:

– the temperature variation in the exchanger does not exceed 2°C;

– the slurry of the body must, according to the requirements, be renewed in a variable between 1 and 4 mn.

The crystallizer is supplied by means of the pump's aspiration.

In a CFCC, the crystal growth essentially occurs in the crystallizer body in which the residence time is predominant. On the other hand, secondary nucleation occurs by attrition in the pump.

The slurry's entry into the body can be:

1) tangential entry (Figure 4.7). The slurry is introduced under the free surface in order to avoid sudden vaporization spurts that can be detrimental to the installation's mechanical integrity. This system is only used in case of vaporization;

2) axial entry (Figure 4.8). This system is used in the case of cooling and also in the case of mechanical recompression of condensation. We will see its advantages.

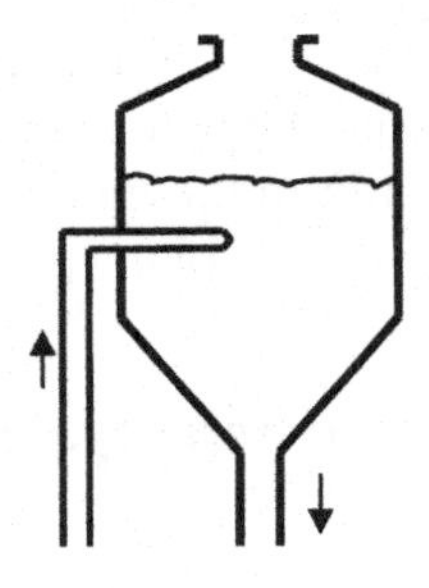

Figure 4.7. *Tangential entry*

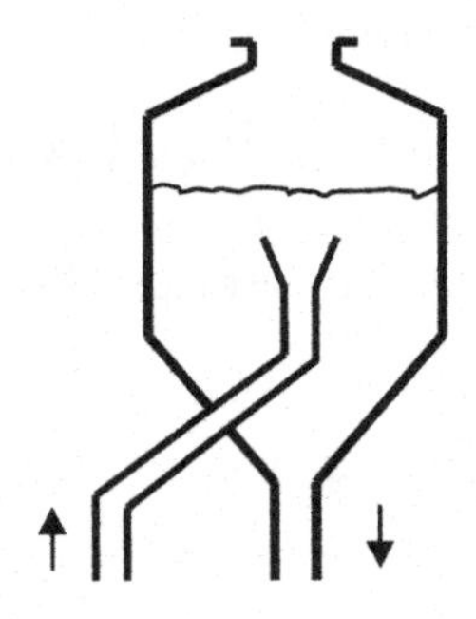

Figure 4.8. *Axial entry*

The circulation flow in a CFCC is high and we can consider that the slurry flow does not vary, but remains equal to Q_B between the body entry and exit, even if there is evaporation. Similarly, the crystal surface A_C does not vary.

In the case of cooling, the liquor exits from the exchanger super-cooled and supersaturated and, if the installation is correctly designed, the loss of supersaturation in the pipes leading to the body is low (the transit time is between 2 and 3 s and does not exceed the metastable limit).

If we proceed by evaporation, the liquor exits the exchanger reheated and under hydrostatic pressure. As it rises to join the crystallizer, the pressure decreases, boiling occurs and the temperature decreases, approaching that of the body if the solubility increases slightly with temperature. On exiting the exchanger, the liquor is reheated, but ebullition occurs, and the liquor cools and becomes concentrated. It becomes supersaturated in the body where the greater part of ebullition occurs.

Between body exit and entry into the exchanger, secondary nucleation can occur by attrition in the pump.

4.3.2. *Residence time of crystals in the installation*

If we use one loop, the various residence times are as follows:

– crystallizer body: 1–4 mn;

– loop auxiliaries (velocity 2 $m.s^{-1}$) and particularly:

- pipes between the exchanger and the body: <3 s,
- exchanger: 2–6 s.

We observe that, as expected, the residence time in the auxiliaries is short relative to the residence time in the body. The characteristics of the latter are determinant here, and the chief among these is volume.

For tangential entry, experience shows that the residence time of crystals is identical to that of the mother liquor if their size does not exceed 40 μm. For larger sizes, crystals do not follow the fluid streams and their residence time in the installation depends on their size:

$$\frac{\tau_C(L)}{\tau_L} = \frac{1}{1 + k L^2} \qquad \text{with } k = 0 \text{ for } L < 40 \mu\text{m}$$

Segregation occurs inside the crystallizer. There is no segregation in the auxiliaries since the flow here is of the piston type at a high velocity.

In general, we can consider that the k coefficient is proportional to $\Delta\rho = \rho_C - \rho_L$

With regard to the axial entry, the larger crystals are projected upwards (within the liquor), taking time to descend so that the k coefficient is lower or even zero.

4.3.3. *Entry level in a vaporization body*

When the level of slurry within the body is greater than 50 cm or more at the level of the entry tube, a fraction of the arriving slurry does not reach the zone above the tube. This fraction of slurry immediately descends into the exit cone without being freed of all the heat corresponding to its superheating. This is followed by an increase in the circulating slurry temperature and a decrease in the exchanger's LMTD (logarithmic mean temperature difference). The thermal transfer power of the exchanger is impaired as a result.

However, we cannot have the tube arrive too close to the slurry surface if we wish to avoid vapor spurts, deposits and encrustation on the walls of the gas space.

Thermal short-circuiting is minimized if we choose axial entry.

4.4. Crystallizer with fluidized bed

4.4.1. *Presentation*

The crystallizer body is a cylinder or a cone with its apex facing downwards, in which the fluid bed made up of crystals and the rising liquid is situated. Crystals descend slowly as they grow, being captured in the lower part. The supply of supersaturated liquor is ensured by a tube that descends to the lower part of the body. Used liquor (that has lost its supersaturation) is decanted by flooding to the surface of the fluidized bed.

This device is suitable for fragile crystals as agitation is very low within the fluidized bed. It is also suited to the production of large crystals by means of a modest production. Indeed, the liquor's rate of climb is low and the productivity connected with the horizontal part of the body varies between 30 and 150 kg.h^{-1}.m^{-2} for crystals in the order of millimeters.

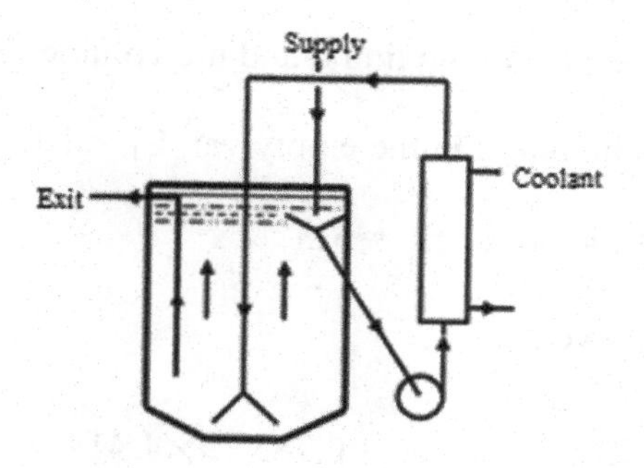

Figure 4.9. *A crystallizer with a fluidized bed*

Secondary nucleation is zero as it is the clean liquor that circulates through the exchanger and the pump. Moreover, supersaturation is maintained just above the metastable limit to provoke a slight nucleation.

Seeds form throughout the bed but particularly in the lower part in which supersaturation ΔX is maximum. While they are climbing to the upper part of the bed, they grow and, having reached a size limit, they descend into the bed where they are decanted at the base.

4.4.2. *Suspension and nucleation power*

The hydrostatic pressure exerted by the crystals at the base of the bed on the liquid is:

$$\Delta P = (1 - \varepsilon)\rho_S gh$$

Assuming the following mean values:

void fraction of bed: $\varepsilon = 0.85$

crystal density: $\rho_S = 1{,}500$ kg.m^{-3}

gravity field: $g = 9.81\ m.s^{-2}$

bed height: $h = 2\ m$

$$\Delta P = (1 - 0.85) \times 1{,}500 \times 9.81 \times 2 = 4{,}414\ Pa$$

Assuming, with respect to the liquid and the volume of the body:

– the velocity of the liquid in the empty vat, $U_L = 10^{-2}\ m.s^{-1}$;

– the diameter of the body, $D_C = 3 m$.

The suspension power is:

$$P_S = U_L \times \frac{\pi}{4} \times D_c^2 \times \Delta P = 10^{-2} \times 0.785 \times 9 \times 4{,}414$$

$$P_S = 312\ W$$

The volume of the body is:

$$\frac{\pi}{4} \times D_c^2 \times h = 0.785 \times 9 \times 2 = 14 m^3$$

Hence, 312/14 = 23 $W.m^{-3}$ for the suspension volume power.

The suspension volume power for a CFC is considerably greater: 200–400 $W.m^{-3}$.

Thus, the secondary nucleation is negligible and the primary nucleation is low.

4.4.3. *Calculation procedure*

Here, we base our method on that proposed by Miller and Seaman [MIL 47].

We estimate the porosity ε, giving it a constant value that is not unreasonable. At the top of the bed, the crystals are small and their slip velocity is low. At the bottom of the bed, the number of crystals is the same but these crystals are larger and their slip velocity is higher. It follows that porosity ε is the result of two values of opposite effect, L_P and V_{gl}.

The empty vat velocity U of the liquid varies between certain limits according to the product treated:

$$3.10^{-3}\,m.s^{-1} < U < 5.10^{-2}\,m.s^{-1}$$

On the other hand, if we accept:

Void fraction of bed: $\varepsilon = 0.8$

Real density of solid: $\rho_S = 1,500\,kg.m^{-3}$

Maximum mass flowrate density: $150\,kg.m^{-2}.h^{-1}$

Fall speed of crystals relative to the workshop is:

$$\frac{150}{1,500 \times 3,600 \times (1-0.8)} = 0.13.10^{-3}\,m.s^{-1}$$

Therefore, we can disregard the velocity of crystals relative to that of the liquid. Thus, assuming immobile crystals, the velocity of the liquid in the empty vat is:

$$U = V_{gl}\varepsilon = V_{gl}\left(1 - n\alpha L^3\right)$$

V_{gl} : slip velocity of the liquid relative to immobile crystals ($m.s^{-1}$)

n: number of crystals per cubic meter of the slurry (m^{-3})

L: size of crystals at the level in question (m).

From the above, we can deduce that:

$$n = \frac{V_{gl} - U}{\alpha L^3 \times V_{gl}}$$

The crystal surface per cubic meter of the bed is:

$$A_c = \beta n L^2 = \frac{\beta}{\alpha L}\left(\frac{V_{gl} - U}{V_{gl}}\right)$$

The growth kinetic is written as:

$$C_E \times \frac{d\Delta X}{d\tau} = -k_G A_c \Delta X^g$$

C_E : mass of free water per cubic meter of the bed (kg.m^{-3})

A_C : crystal area per cubic meter of the bed (m^{-1}).

We can write out:

$$\frac{d\Delta X}{d\tau} = \frac{d\Delta X}{dx} \times V_{gl}$$

By knowing the desired production P (in kg/h), together with the desired product size L_P, we can deduce the product JV of the nucleation rate J by the volume V of the bed:

$$JV = \frac{P}{3{,}600\,\alpha \rho_s L_p^3}$$

We establish the material balance for supersaturation to disappear completely through a fluidized bed:

$$W_E\, \Delta X_P = J\, V\, \alpha \rho_s L_p^3$$

W_E : free water flow (kg.s^{-1})

ΔX_P : supersaturation at the bed base.

From this equation, we deduce the value of W_E corresponding to ΔX_p. In addition, on a certain level:

$$W_E \, \Delta X = J\,V\,\alpha\rho_s\,L^3$$

Differentiating relative to height z in the bed:

$$W_E \frac{d\Delta X}{dz} = 3JV\alpha\rho_s L^2 \frac{dL}{dz}$$

If we replace $\frac{d\Delta X}{dx}$ and ΔX in the kinetic equation with their values, we obtain:

$$C_{EO} \frac{JV}{W_E} 3\,\alpha\rho_s \frac{dL}{dz} = -k_G \frac{\beta}{\alpha}\left(\frac{V_{gl} - U}{V_{gl}^2}\right)\left(\frac{J\,V\,\alpha\rho_s}{W_E}\right)^g L^{3(g-1)}$$

Write:

$$a = C_{EO} \frac{JV}{W_E} 3\,\alpha\rho_s$$

$$b = k_G \frac{\beta}{\alpha}\left(\frac{J\,V\,\alpha\rho_s}{W_E}\right)^g$$

The kinetic equation becomes:

$$a\frac{dL}{dz} = -b\left(\frac{V_{gl} - U}{V_{gl}^2}\right)L^{3(g-1)}$$

We integrate this equation in order to have the profile of L and that of ΔX.

In the calculation of a, we see that C_{EO} intervenes, which is calculated as follows:

$$C_{EO} = \varepsilon \rho_L \times \frac{1}{1+X_o}$$

ε : fraction of fluid volume in the bed

ρ_L : liquor density ($kg.m^{-3}$)

X_o : solute ratio (kilogram of solute per kilogram of liquor).

EXAMPLE.–

Consider the crystallization of $K_2 SO_4$ with the following data (taken from an example treated differently by Mullin [MUL 72]).

$P = 1{,}000\ kg.h^{-1}$	$\alpha = 1$	$K_G = 10$
$\rho_S = 2{,}660\ kg.m^{-3}$	$\Delta X_P = 0.04$	$\beta = 6$
$L_\rho = 10^{-3}\ m$	$\rho_L = 1{,}082\ kg.m^{-3}$	$U = 0.04\ m.s^{-1}$
	$X_0 = 0.15$	$g = 2$

$$JV = \frac{1{,}000}{3{,}600 \times 1 \times 2{,}660 \times 10^{-9}} = 1.04.10^5\ s^{-1}$$

$$W_E = \frac{1 \times 2{,}660 \times 10^{-9} \times 1.04.10^5}{0.04} = 6.9\ kg.s^{-1}$$

Assume that, at the base of the bed, the volume fraction of liquid is:

$$\varepsilon = 0.8$$

So:

$$V_{gl} = \frac{0.04}{0.8} = 0.05\ m.s^{-1}$$

$$V_{gl} = 0.04 / 0.8 = 0.05 \text{ m/s}$$

$$C_E = 0.8 \times 1,082 \times \frac{1}{1+0.15} = 753 \text{ kg.m}^{-3}$$

$$a = \frac{753 \times 1.04.10^5 \times 3 \times 1 \times 2,660}{6.9} = 9.10^{10}$$

$$b = 10 \times 6 \left(\frac{1.04.10^5 \times 1 \times 2,660}{6.9} \right)^2 = 9.6.10^{16}$$

The kinetic equation is written:

$$9.10^{10} \frac{dL}{dx} = -9.6.10^{16} \left(\frac{0.05 - 0.04}{0.05^2} \right) L^3$$

So:

$$\frac{dL}{L^3} = -4.26.10^{16} dx$$

We will call L_o the size of crystals at height x that will be defined as the bed summit. After integration, we obtain, for a height of 6 m, the size limit at the bed summit:

$$\frac{1}{L_o^2} - 10^6 = 2 \times 4.26.10^{16} \times 6$$

$$L_o = 1.2.10^{-4} \text{ m}$$

4.4.4. *Laboratory study of a fluidized bed*

We use a mass M of crystals that all have a size equal to the size limit set for the crystallizer summit. The crystals are fluidized with the saturated liquor at empty vat velocity U.

The bed cross-section is A and its measured height is H. Its volume is, accordingly, AH.

If ρ_s is the density of the crystals:

$$\varepsilon = \frac{AH - M/\rho_s}{AH} = 1 - \frac{M}{\rho_s AH}$$

According to Richardson and Zaki:

$$U = V_l \varepsilon^n$$

Or rather:

$$\text{Ln } U = \text{Ln } V_l + n \text{ Ln } \varepsilon$$

The straight line obtained in logarithmic coordinates provides V_ℓ and n.

We know that:

$$V_{gl} = \frac{U}{\varepsilon} = V_l \varepsilon^{n-1}$$

V_ℓ : limit fall velocity of an isolated particle of size L ($m.s^{-1}$)

n: exponent function of V_ℓ (see section 4.2.4).

We can write out:

$$\varepsilon = \left(\frac{U}{V_l}\right)^{1/n}$$

and:

$$V_{gl} = V_l^{1/n} U^{1-1/n}$$

4.5. Elutriation columns

4.5.1. *General*

These columns act to free the crystals of interstitial fines, they are often arranged at the base of a crystallizer.

Large crystals descend into the column while the liquor rises against the current, carrying the fines away, which provides a certain classification of crystals. Once the fines have been removed, the large crystals are decanted at the foot of the column. Furthermore, these columns can be used to cool crystals by means of the rising liquor.

If we are operating with an undersaturated liquor, we must dissolve the fines and thus increase the mean crystal size. On the other hand, if we are operating with a supersaturated liquor, we expel the fines in the crystallizer and reduce the mean crystal size, which is undesirable.

The phenomenon of fluidization applies in the operation of these columns.

4.5.2. *Residence time of crystals in the column*

Where H_E is the column height, along this height, the mass of crystals per m² of cross-section is:

$$M_C = H_E \; D_B \rho_B$$

D_B : slurry "density": here, crystal mass divided by slurry mass

ρ_B : slurry density (kg.m^{-3}).

If we provide ourselves with the elutriation flow G_E, that is, the flow density in the crystal mass crossing a unit of column cross-section, we immediately obtain the mean residence time of crystals in the column.

$$\tau = \frac{H_E \, D_B \, \rho_B}{G_E}$$

EXAMPLE.–

$H_E = 5 \text{ m} \quad D_B = 0.4 \quad \rho_B = 1{,}300 \text{ kg.m}^{-3} \quad G_E = 4 \text{kg.m}^{-2}.\text{s}^{-1}$

$$\tau = \frac{5 \times 0.4 \times 1{,}300}{4} = 650 \text{ s} = 11 \text{ mn}$$

This time is enough to obtain the classification of solid particles in a bed fluidized by a fluid and whose solid particles vary in size.

4.5.3. *Calculation of different liquid flows*

For this, we will use the drawing of a column as represented in Figure 4.10.

Decanted slurry is of the same composition as the slurry located above the lateral outlet. Accordingly, we can outlet the porosity of this slurry:

$$\varepsilon = \frac{(1-D_B)/\rho_L}{\frac{D_B}{\rho_S} + \frac{1-D_B}{\rho_L}}$$

If we provide ourselves with flow W_S of crystals together with the elutriation flow rate density G_E, we can immediately deduce cross-section A_E of the column.

$$A_E = W_S / G_E \quad \text{hence the diameter } \phi = \sqrt{\frac{A_E}{0.785}}$$

The liquid flow accompanying the crystals is:

$$Q_{LP} = \frac{W_S(1-D_B)}{\rho_L D_B}$$

Now, let us place ourselves at a given location above the lateral outlet.

If the liquid were immobile, the crystals' fall velocity relative to the workshop would be, according to Richardson and Zaki's law:

$$V_{\ell 1}\varepsilon^n$$

$V_{\ell 1}\varepsilon^n$: fall velocity limit of a particle in the dispersion (m.s^{-1}).

However, the liquid rises with an empty vat velocity V_L, which would slow the crystals, which drop to velocity V_C relative to the workshop.

$$V_C = V_{\ell I}\varepsilon^n - V_L$$

In addition, the flow density of the crystals is:

$$G_E = (1-\varepsilon) V_C \rho_S$$

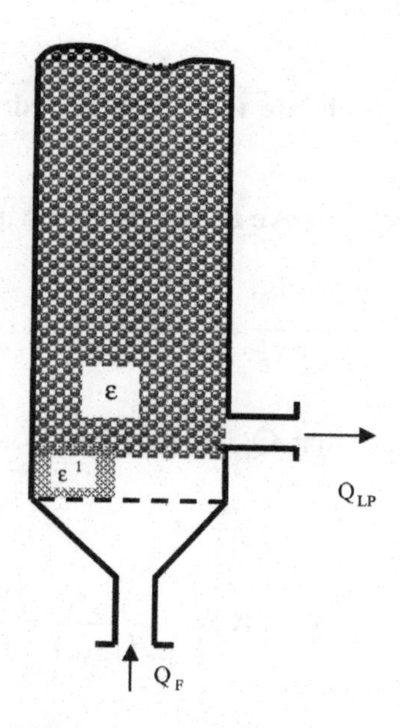

Figure 4.10. *Schematic diagram of an elutriation column*

so:

$$G_E = (1-\varepsilon)\rho_S \left(V_{\ell I}\varepsilon^n - V_L\right)$$

In this expression, everything is known except V_L, which can be calculated as follows:

$$V_L = V_{\ell I}\varepsilon^n - \frac{G_E}{(1-\varepsilon)\rho_S}$$

The flow rate entering the crystallizer is:

$$Q_L = V_L A_E$$

We can now calculate the liquid flow rate Q_F fed at the bottom of the column:

$$Q_F = Q_L + Q_{LP}$$

NOTE.–

We will show that crystals are more compacted under the lateral outlet than above it.

The ratio ε'/ε is placed relative to 1 in the same manner as:

$$y = \frac{V_{\ell l}(\varepsilon')^n}{V_{\ell l}\,\varepsilon^n} = \frac{Q_F}{(V_C + V_L)A_E} = \frac{Q_L + Q_{LP}}{Q_L + V_C A_E} < 1$$

It should suffice to show that Q_{LP} is less than $V_C A_E$.

Indeed:

$$Q_{LP} = \frac{W_S}{\rho_L}\frac{(1-D_B)}{D_B} \text{ and } V_C A_E = \frac{W_S}{\rho_S(1-\varepsilon)} = \frac{W_S}{D_B}\left(\frac{D_B}{\rho_S} + \frac{1-D_B}{\rho_L}\right)$$

so:

$$\frac{1-D_B}{\rho_L} < \frac{(1-D_B)}{\rho_L} + \frac{D_B}{\rho_S}$$

which is evident.

The lower part, in which crystals are immobile and compressed, is of no interest here. It is therefore recommended to place the liquid injection grid directly above the lateral tube.

EXAMPLE.–

$$D_B = 0.4 \qquad \rho_S = 2,200 \text{ kg.m}^{-3}$$

$$W_S = 1.1 \text{ kg.s}^{-1} \qquad \rho_L = 1,200 \text{ kg.m}^{-3}$$

$$G_E = 4 \text{ kg.m}^{-2}.\text{s}^{-1} \qquad V_{\ell l} = 0.56 \text{m.s}^{-1} \left(d_P = 10^{-3} \text{m}\right)$$

$$n = 3.408$$

We calculate:

$$\varepsilon = \frac{0.6/1200}{\dfrac{0.6}{1,200} + \dfrac{0.4}{2,200}} = 0.735$$

$$A_E = 1.1/4 = 0.275 \text{ m}^2$$

so $\phi = 0.6$ m

$$Q_{LP} = \frac{1.1 \times (1-0.4)}{1,200 \times 0.4} = 1.375.10^{-3} \text{ m}^3.\text{s}^{-1}$$

$$Q_L = 0.275\left(0.056 \times 0.735^{3.408} - \frac{4}{(1-0.735)\,2,200}\right)$$

$$Q_L = 3.6.10^{-3} \text{m}^3.\text{s}^{-1}$$

$$Q_F = 1.375.10^{-3} + 3.6.10^{-3} = 4.98.10^{-3} \text{m}^3.\text{s}^{-1}$$

so $\quad Q_F = 17.9 \text{m}^3.\text{h}^{-1}$

NOTE.–

Flow rate density G_E has been taken as equal to 4 kg.m^{-2}.s^{-1} for particles whose size is of the order of millimeters, but we must adopt a lower value for significantly smaller crystals to avoid a negative value of V_L, which would mean that the column is a simple sedimentation column without any elutriation.

4.6. Crystallizer piston with scraped walls [ULL 86]

4.6.1. *Description*

The device is a horizontal cylinder fitted with helical scrapers. It is equipped with a double envelope. The force of pressure causes the crystals and the liquor to progress in the same direction.

The coolant circulates counter-current to the suspension in the envelope.

The exchange surface is in the order of 1 m^2 per linear m. The scrapers turn slowly (5–20 rev. mn^{-1}). The calculation and experience show that the transfer coefficient varies from 100 to 300 $W.m^{-2}.°C^{-1}$.

Small and relatively uniform crystals can also be obtained.

The production of a tube is in the order of a dozen tons per day. Note that the device can easily accept a difference in temperature of 15°C between the slurry and the coolant.

4.6.2. *Calculation elements*

The equations that were established for the cooled crystallizing vat also apply here by replacing $\Delta\tau$ by $\Delta x / V$, where V is the slurry's rate of progression.

However, the movement of the scrapers and the presence of crystals lead to an axial dispersion within the liquid, and rather than writing as for the cooled crystallizing vat:

$$C_E\, X + M_C = X_o\, C_E$$

we will write:

$$C_E \left(X - K \frac{dX}{dx} \right) + M_C = X_o C_E$$

We will make our first calculation neglecting the axial dispersion. From the profile obtained in this way for X, we will iterate on the complete equation. During the first calculation, we assume a linear temperature profile for the coolant.

4.7. Batch crystallizing: homogeneous vat

4.7.1. *Design of batch crystallizers*

We must make the distinction between:

– crystallizers with solvent vaporization;

– cooled crystallizers.

1) Vaporization crystallizers:

In theory, it is possible to immerse the heating surface in the crystallizer even when agitated. In order to avoid boiling on the wall provoking fast crusting of the tubes and a drop in the system's thermal power, we must provide a liquid height of 50 cm above the tubes (which are vertical).

Another solution consists of providing an external exchanger, in which we can circulate the slurry at a high velocity (2 $m.s^{-1}$). To avoid boiling on the exchanger's wall, we can provide a sufficient hydrostatic height rather than a release valve on the return pipe between the exchanger and the crystallizer.

2) Cooled crystallizer:

We can avoid using the external exchanger and circulation pump by performing the thermal transfer using an immersed coil, which is less expensive. Indeed, the classic coil fits the relationship:

$$\frac{\text{Transfer surface}}{\text{Crystallizer volume}}$$

when the crystallizer's volume does not exceed 5 m^3.

When we introduce an undersaturated load, any crusting on the coil is dissolved.

We can use a draft tube fitted with a marine impeller and place the coil in the annular space. This arrangement guarantees high circulation. We must ensure that the temperature difference between the slurry and the cooling fluid is sufficiently low to void crusting. High levels of recirculation may be required.

Crystallizing vats were studied in detail by Wey and Estrin [WEY 73], by Jones [JON 74a] and by Jones and Mullin [JON 74b].

4.7.2. *Batch crystallization methods (presentation)*

The aim is to create the initial distribution of crystals, which will then be grown.

If, at a given time, supersaturation happens to be too high, there can be a massive primary nucleation that spreads the product's size distribution, decreasing its mean size.

Accordingly, supersaturation must be controlled. One method for this consists of using a supersaturation program, whereby the product's granulometry should be close to a given granulometry. However, this method, proposed by Jones [JON 74a], is mathematically complex, and is incongruous in industrial practice.

On the other hand, predictive calculations were performed by Jones and Mullin [JON 74b]. These calculations, assisted by practical experience, showed that an operation performed at constant supersaturation provides larger crystal sizes than a "natural" cooling and narrower size distribution. Let us consider this in more detail:

1) "Natural" cooling or vaporization (not controlled).

The thermal power, initially strong, decreases exponentially. Supersaturation, initially high, leads to the formation of numerous seeds over a non-negligible time. The seeds are numerous and their size distribution is spread.

Thus, the product's quality depends on the thermal power. The greater the power, the smaller the crystals and the greater the granulometry spread due to the high initial nucleation.

2) Constant supersaturation.

This method was recommended to narrow the granulometry and increase the crystal size. On first impression, if the supersaturation is constant, we can suppose that growth G and nucleation velocity J are also constant. According to this hypothesis, the thermal power increases considerably from the beginning to the end of the operation. Indeed, we require 100 times more energy to increase the size of a crystal of 1 mm by 20 μm than to increase a crystal of 100 μm by the same amount. It would be difficult to apply such a method in industry.

In practice, for an effective ("controlled") operation, we use neither of these methods.

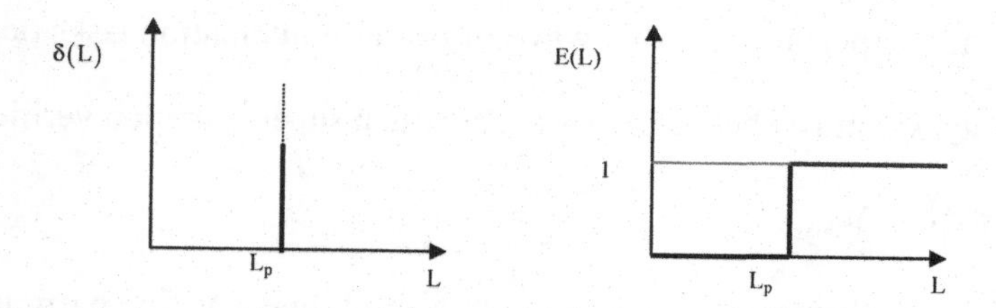

Figure 4.11. *Ideal controlled crystallization*

3) Controlled process.

The purpose of batch crystallization is to obtain a narrow crystal centered on a single size. In this case, the population density presents a sharp spike for the desired crystal size.

To allow for a brief mathematical parenthesis, we will say that the P.D (population density) is represented by a Dirac impulse function $\delta(L_P)$. A better term would be Dirac "distribution".

This function's integral is the echelon function E(x).

$$\int_0^\infty n(L)dL = \int_0^\infty \delta(L_P)dL = E(L_P) = \text{number of crystals}$$

and the P.D. with an integral equal to 1 must be:

$$n(L) = \delta(L_n)$$

This objective can be approached in two ways: by seeding or by controlled supersaturation.

4.7.3. *Seeding*

After milling and/or drying, the seed must be washed and lightly dissolved to eliminate dust.

It is not easy to add the seed at exactly the moment that the mother liquor is saturated. If we act:

– too early (undersaturation), the seed is dissolved;

– too late (supersaturation), a wave of primary nucleation risks occur.

In reality, we must choose the instant at which supersaturation verifies Δc.

$$0 < \Delta c < \Delta c_{max}$$

There is an advantage in seeding with a small mass of fine particles.

If there is not attrition, we accept that the number of crystals present in the crystallizer is invariable in time and equal to the number of seeds dispersed in the device at the beginning of the operation, but there is a dependence relationship between the number and the size of seeds, on the one hand, and the cooling or vaporization rate, on the other hand.

If supersaturation is constant, so is the growth velocity G. At a given moment, the apparition rate of the crystal mass is:

$$\frac{dM_c}{dt} = A_c G$$

A_c is the crystal surface, all the same size ($L_s + Gt$):

$$A_c = N_s \beta (L_s + Gt)^2$$

N_s is the number of seeds that we assume to grow at the same rate:

$$N_s = \frac{M_s}{\alpha L_s^3}$$

M_s and L_s are the total mass and the seed dimension.

Hence:

$$\frac{dM_c}{dt} = \frac{M_s}{\alpha L_s^3} \beta G (L_s + Gt)^2 \qquad [4.4]$$

On the other hand, if X is the mass of solute per kilogram of the solvent, we can write:

Cooling $$\frac{dM_c}{dt} = -\frac{dX^*}{dT}\frac{dT}{dt} M_{solvent}$$

Vaporization $$\frac{dM_c}{dt} = -X^* \frac{dM_{solvent}}{dT}$$

Hence, the relationship between M_s and L_s, and dX^* / dT or $dM_{solvent} / dt$.

We observe that if dM_c / dt is maintained constant, the growth G varies, thus:

$$G \propto (1 + Gt / L_s)^{-2}$$

This is not easy to obtain, since we have hypothesized that the supersaturation was constant. If we adhere to this last hypothesis, we must necessarily alter dM_c/dt throughout the crystallization process. This equation is readily integrated and the crystal mass is accordingly:

$$M_c = \frac{M_s \beta}{3\alpha L_s^3}(L_s + Gt)^3 = M_s\left(1 + \frac{Gt}{L_s}\right)^3$$

We observe in passing that $\beta = 3\alpha$, if the crystals, while growing, remain similar to themselves.

4.7.4. *Controlled supersaturation*

We must, from the beginning, dispose of a precise number N_o of seeds. Therefore, if L is the intended size of the crystals and M is the crystals' intended mass, then, in these conditions:

$$N_o = \frac{M}{\alpha L^3 \rho_c}$$

The seeds can result from:

1) seeding with N_o seeds of size L_o, so with mass $N_o \alpha L_o^3 \rho_c$;

2) an initial primary nucleation of duration t.

$$N_o = \int_0^t J(t)\,dt \qquad \text{with } J = J(\Delta c) \text{ and } \Delta c = \Delta c\ (t)$$

The supersaturation over this initial period is:

$$\Delta X = X^* - X \qquad \frac{d(\partial X)}{dt} = \left(\frac{\partial X^*}{\partial t}\right) dt \qquad \frac{dM_e}{dt} = \frac{Q}{r} \qquad X^* = \frac{M_s}{M_e} = \text{cste}$$

Cooling $\left(X^* \text{ vary}\right)$ — Vaporization $\left(M_e \text{ vary}\right)$

Q: thermal power (W)

r: vaporization heat ($J.\ kg^{-1}$).

The crystal mass is:

$$M_c = \rho_c \alpha \int_0^t J(\tau)\left[G(t-\tau) + L_0\right]^3 d\tau \qquad \tau < t$$

At time t, the number of crystals $J d\tau$ have appeared at time τ and have grown over time $(t-\tau)$.

Hence:

$$\frac{dM_c}{dt} = 3\alpha\rho_c G \int_0^t J(\tau)\left[(t-\tau)G + L_0\right]^2 d\tau$$

J is a function of $\Delta X = X - X^*$ with $X = M_s / M_e$ and $dM_s = -dM_e$.

Cooling $$\Delta X = \frac{M_S - M_S^*}{M_e} \rightarrow M_e \frac{d(\Delta X)}{dt} = -\frac{dM_c}{dt} - \frac{\partial M_S^*}{\partial \theta} \times \frac{d\theta}{dt}$$

Vaporization $$\Delta X = \frac{M_S - M_S^*}{M_e} \rightarrow \frac{d(\Delta X)}{dt} = d\left(\frac{M_S}{M_e}\right) = \frac{1}{M_e}\frac{dM_S}{dt} + \frac{M_S}{M_e^L}\frac{Q}{r_e}$$

$(M_s^* = \text{const.})$

$$\Delta X(t) = \int_0^t \frac{d(\Delta X)}{d\tau} d\tau$$

The logical sequence of the calculations is as follows:

$$\Delta t \rightarrow \Delta M_c \rightarrow \Delta(\Delta X) \rightarrow J(t + \Delta t)$$

In other words:

$$\frac{dJ}{dt} = f(J, t)$$

The Runge–Kutta method of order 4 is suitable to study nucleation variations according to time (see Appendix 1).

In fact, $d\theta/dt$ results in a decrease when the temperature θ decreases (see exchanger theory). Similarly, power Q also decreases.

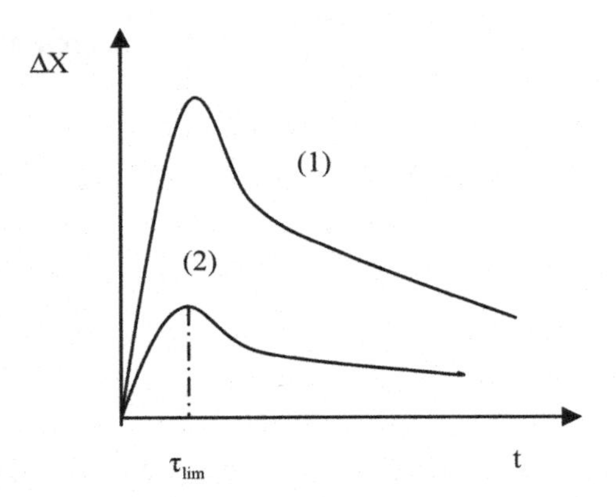

Figure 4.12. *(1) "Natural" program; (2) controlled program*

Finally, ΔX decreases due to the rapid increase in dM_c/dt with t, so that J nucleation reaches a maximum before decreasing until ΔX, having a moderate value, is only of use to crystal growth.

After this instant, $d(\Delta X/dt)$ moves towards zero, which provides the desired values for dT/dt and Q. The τ parameter, which is the instant of crystal formation, has reached limit τ_{lim}, so that, for $t > \tau_{lim}$, the J nucleation is very low.

According to Figure 4.12 in the case of non-seeded nucleation, supersaturation control leads to a reduction in the spike amplitude of primary nucleation.

Similarly, crystallization after seeding leads to a variation in the neighboring supersaturation of curve (2), for which the supersaturation maximum has been weakened significantly.

4.8. Vat population density

4.8.1. *Constant nucleation without growth*

The number of particles increases in proportion to time.

$$N = Jt \qquad (m^{-3})$$

4.8.2. *Growth without nucleation*

$$t = 0 \rightarrow n = n_o(L_o) \text{ absolutely arbitrary} \qquad \forall L_o$$

Function n_o does not change by L that goes from L_o to $L_o + Gt$

$$\text{Ln}n_o = f(L_o) \qquad \text{Ln}(n_o) = f(L - Gt) = f(L_o)$$

The number of crystals does not vary:

$$N = \int_{Gt+L_o}^{\infty} n(L,t)\,dL = \text{cste} \qquad \forall t$$

4.8.3. *Nucleation with growth*

The partial derivative equation, with n as the solution, is:

$$\frac{\partial n}{\partial t} + G\frac{\partial n}{\partial L} = \frac{J}{L_o}$$

A possible solution is:

$$n = n_o \exp\left(\frac{Gt - L}{L^*}\right) + \frac{Jt}{L_o}$$

At the beginning of this operation, $t \# 0$ and we obtain:

$$n = n_o \exp\left(-\frac{L}{L^*}\right)$$

Experience confirms this result that represents the *population density of detectable seeds at the beginning of the operation.*

4.9. Choice of crystallization

4.9.1. *Cooling or vaporization*

The process by cooling is required if the relationship between temperature and concentration to saturation is such that:

$$\frac{d\left(Lnc^{*}\right)}{d\left(LnT\right)} > 8$$

Vaporization is required *a priori* if:

$$\frac{d\left(Lnc^{*}\right)}{d\left(LnT\right)} < 2$$

Thus, the solubility of sodium chloride is practically independent of temperature, as that of K_2CO_3 and $(NH_4)_2CO_3$. However, many crystallizations occur by vaporization for another reason. In fact, vaporization allows us to reach yields approaching 1 (if the purge is low), which is not the case for cooling.

Cooling a solution leads to a mediocre depletion of the mother liquor. If we cool too much, we risk crystallizing a more hydrated crystal species than desired and, in extreme cases, crystallizing the water itself.

If we want to obtain a crystal species that is greatly hydrated, the evaporation technique requires low temperature, that is, a high vacuum, and occasionally, we are obliged to proceed by cooling anyway.

In order to crystallize by cooling, we must ensure that solubility varies in a significant way with temperature, although this is not a sufficient condition and, very often, we crystallize such products by evaporation for the reasons of yield addressed above.

4.9.2. *Continuous operation or batch operation*

There are two essential criteria:

– the production capacity;

– the quantity of product.

1) Capacity:

In continuous operation, it is difficult to descend below 1 $m^3.h^{-1}$ of dry powder. Indeed, blockages become constant when the slurry extraction tube has an internal diameter less than or equal to 2 cm, which corresponds to 2 $m^3.h^{-1}$ of slurry at 2 $m.s^{-1}$ and, if the volume fraction of crystals is 0.2 and the void fraction of dry powder 0.5, the minimum production flow of powder is:

$$\frac{2 \times 0.2}{0.5} = 0.8 \text{ m}^3.\text{h}^{-1}$$

In practice, it is wise not to descend below 1 $m^3.h^{-1}$ of dry powder.

The advantage of the continuous process is pronounced when the capacity increases in terms of economy, as the costs of studies and research together with monitoring are independent of the capacity. Typically, we rarely exceed 20 $m^3.h^{-1}$ of dry powder.

2) Product quality:

The batch method is the most flexible, and by correctly programming the temperature or the vaporization power, we can modulate granulometry. On the other hand, the product's quality can vary from one change to another. We must also remember that seeding a vat at the beginning of the operation imposes the crystal number and avoids the uncertain process of primary nucleation.

4.9.3. *Agitation and thermal exchange*

If the requirement of exterior heat is significant or rather, if the residence time is short enough for a low crystallizer volume, the crystallizer can be agitated correctly by the exchanger's exterior loop arrival current, as the recirculation flow will be significant. This is the principle of *forced circulation crystallizers*.

On the other hand, *if the heat requirement is low* or the body's volume significant, the exchanger's loop flow will be reduced and will not agitate the device. *We must then provide for internal agitation*. Indeed, the study of solid suspensions shows that, in order to obtain a correct homogeneity, we need to create the rapid circulation from top to bottom and from bottom to top within the device. This reduces the propensity to decantation due to gravity. This circulation can be ensured by a draft tube fitted with an impeller.

4.9.4. *Fragile crystals*

The fluidized bed is well-suited to fragile crystals that are sensitive to attrition. In addition, the fluidized bed is itself a device for product classification. This occurs via the clean liquor circulating in the exchanger of the external loop. There is consequently no attrition in the pump.

Owing to the classification, the crystals collected at the crystallizer base have an increased mean size. This sort of device is often cone-shaped with the apex pointed downwards, which intensifies the classification.

However, the fluidized bed's productivity is mediocre, since in order to obtain the appropriate flow of clean liquor, a large horizontal surface is required.

4.9.5. *Very thick slurry*

When the system is such that the crystal volume (including the interstitial voids) exceeds 30% of the slurry volume, the slurry flows poorly. These are typically crystals in plates or fibers, whose porosity at rest exceeds 80%. Not much liquid remains free to ensure the flow of slurry, leading to considerable attrition.

In such cases, we use a continuous cooled tube crystallizer fitted with internal helical scrapers and a double envelope in which the cooling liquid circulates.

In terms of crystallization, the process is comparable to that of a 'discontinuous' batch, and whereas the residence time of the vat is in the order of several hours (8–24 h for sugar), it is only in the order of a dozen minutes for the cooled tube. The crystals cannot grow much.

4.9.6. *Granulometry spread*

In theory, it is batch crystallization that allows for the greatest control, particularly if prior seeding has been performed. The crystals are also all the same age and, hopefully, also the same size, and the variation coefficient will be in the order of only 10%.

The variation coefficient of a continuous homogenous crystallizer without attrition is of 50%, but can be reduced to 30% or even 20% if the device is fitted with fines destruction associated with product classification.

The V.C. of a fluidized bed equipped with fines destruction is below 20%.

4.9.7. *Crystal purity*

If the liquor is exhausted faster in the product than in impurities, then batch crystallization is required.

4.9.8. *Cascade of serial crystallizers*

The value of a cascade is twofold compared with a single large device:

– to limit supersaturation in each device;

– to avoid supersaturation instability, which is always possible in a large device.

The concurrent circulation of liquor and slurry has the consequence that, with more devices, we approach more operation of the double envelope tube (in the case of cooling); in other words, we approach batch crystallization. This explains why the median size increases with the number of serial devices.

However, this system includes many pumps, and attrition can be severe.

4.9.9. *Particles in the slurry*

In the liquid of an industrial crystallizer, we can encounter a certain particle diversity:

– real seeds consecutive to primary nucleation;

– apparent seeds resulting from the agglomeration of real seeds;

– fragments of size between 1 and 10 μm resulting from the attrition (wear) of significantly larger crystals;

– fragments of size from 30 to 100 μm resulting from the shattering of crystals of size 100–1,000 μm, following an impact with the pump impeller or agitator impeller.

4.9.10. *Possible crystal dimensions*

Of course, the size obtained for the crystals depends not only on the system in question, but also on the type of crystallizer used:

Crystallizer type	*Median size (mm)*
Forced circulation	0.2–0.5
Draft tube	0.5–2
Fluidized bed (and batch crystallization)	1–10

4.10. Exploitation parameters

4.10.1. *Order of magnitude for parameters*

In order for nucleation to be considerable, we must have σ for the supersaturation:

$$\sigma > 0.5$$

On the other hand, in order for growth to be progressive while remaining economically acceptable, we need:

$$0.01 < \sigma < 0.1$$

In this situation, we typically have:

$$10^{-9}\ \text{m.s}^{-1} < G < 10^{-7}\ \text{m.s}^{-1}$$

In the industry, the size of crystals is: $(L = G\tau)$

$$10^{-5}\,m < G\tau < 10^{-3}\,m$$

The residence time τ, the volume fraction φ_T of crystals in slurry and the agitation power per kilogram of slurry $\bar{\varepsilon}$ all depend on the type of the crystallizer.

	τ (hours)	φ_T	$\bar{\varepsilon}$ (W/kg)
Forced circulation	1–2	≤ 0.15	0.2–0.5
Draft tube	3–4	≤ 0.20	0.1–0.3
Fluidized bed	2–4	≤ 0.30	0.1–0.3 (pump)
Batch	1–20	≤ 0.20	0.1

4.10.2. *Residence time and supersaturation*

We can always develop an installation capable of the thermal power corresponding to the crystallization of a given product.

A further choice presents itself. This is:

– the volume V_c of a continuous crystallizer;

– crystallization time τ for a batch crystallizer.

Both of these parameters set the supersaturation value.

If they are low, the supersaturation in the slurry will be too high for crystallization to follow the thermal transfer. Nucleation is high, as is the proportion of fine crystals. The granulometry is spread and the mean size low; too high a supersaturation leads to crystals of irregular shape and fragile texture. These crystals can feature inclusions of mother liquor and thus be impure.

Inversely, there is a growth limit of V_c or τ. This limit is set by the costs of investment.

Ultimately, the choice of V_c or of τ is the result of a compromise. The crystallizer volumes are the result of the value of residence time in the slurry, which *can vary according to the product from 1 h to 12 h (viscous liquid)*. On the other hand, the length of a batch operation is between 2 and 10 h.

4.10.3. *Crystal content in slurry*

Crystallizer's instability has long been a concern of researchers, but for the most part this is merely a problem in the laboratory. In industry, to avoid the problem, it suffices to regulate the concentration of crystals M_c in the crystallizer slurry.

Let us suppose first that M_c is too low. In the case of a vaporization, it suffices to reduce the slurry decantation, and boiling continues along with crystal growth. The concentration M_c increases and regains its set value. If it is a cooled crystallizer, we can choose the clarification zone, extract the mother liquor and send the thick underflow to the crystallizer.

If M_c is now too high in the case of vaporization, it suffices to increase the slurry outflow. In the case of cooling, we recycle a part of the clear liquor to the crystallizer.

We could be led to believe that, for a cooled crystallizer, both possibilities of action described above (M_c too low or too high) should be provided for in the installation, which would call for a complex valve system. In reality, the set value of M_c is different from the value that we call natural (set by the crystallizer temperature), so that regulation always acts in the same direction but in a more or less intense manner.

If we vaporize it, it should suffice to provide a control valve on the slurry outlet. This valve must be placed on a high point and be a direct passage to avoid becoming blocked.

The choice of set value for M_c must be made carefully.

If M_c is high:

– pipe blockages are frequent;

– incrustations and crusting are a danger;

– the crystals produced are larger as the supersaturation is used to deposit solute on the crystal surface rather than causing seed formation.

In practice, it is more desirable to set φ_T volume fraction than M_c in the slurry:

$$0.15 < \varphi_T < 0.25$$

4.10.4. *Calculation of thermal transfer*

The thermal transfer coefficient calculation for the slurry is performed as if the liquor is alone. This process is cautious as:

– the presence of crystals scrapes and activates the boundary layer;

– crystal conductivity is between 1 and 10 $W.m^{-1}.°C^{-1}$ and is greater than that of the liquid (0.1 at 0.5 $W.m^{-1}.°C^{-1}$).

The thermal transfer coefficients vary from 200 to 2,000 $W.m^{-2}.°C^{-1}$ according to the properties of the liquid phase (thermal conductivity and viscosity). We refer to the calculation method in section 2.2.2 of [DUR 16].

4.10.5. *Agitation and attrition*

Agitation plays a double role:

– to homogenize the suspension, so that, in continuous, non-fluidized crystallizers, the residence time is precise for a given size;

– to homogenize the liquid phase in order to avoid local supersaturations other than the mean supersaturation in the device.

The draft tube system with a marine impeller is the best as it allows for a considerable internal recirculation flow. Attrition in a draft tube is negligible if the clearance between the tube and the impeller is greater than or equal to 2 cm and if the peripheral velocity of the impeller is less than 8 $m.s^{-1}$.

Nonetheless, the effects of attrition are still a danger in the following rare instances:

– the crystals are particularly fragile due to the their jagged form or their weak cohesion;

– the installation includes a variety of slurry pumps (multilevel crystallizers).

Fragments resulting from attrition only come from crystals of a minimum size. Indeed, crystals of a size smaller than 100 μm faithfully follow their fluid streams and do not collide with solid bodies (agitator impeller, pump wheel). The size of fragments resulting from abrasion is in the order of 1–100 μm.

4.10.6. *Incrustations*

A priori, the accumulation of crystals on the walls can occur both on the crystallizer body as well as in the exchanger.

In the body, if blocks of crystals form by splashing, this leads to:

– the agitator becoming unbalanced;

– the incorrect operation of temperature, pressure and density detectors;

– blockages when the blocks detach and enter pipes.

In order to decrease the influence of splashes, we must avoid using rough surfaces, onto which drops of supersaturated liquor can stick. Preferably, the material used should be a smooth or polished surface (rubber or enameled steel).

Inside the exchanger, if an internal tube wall becomes lined with crystals (crusting), the power of the thermal transfer is reduced together with the installation's capacity, and in extreme cases, the tubes may become blocked. For this, there are three categories of solutions:

– avoiding excessive supersaturations. In case of cooling, we must enter the labile zone and, in case of vaporization, boiling on the exchanger walls is to be avoided;

– an internal tube diameter of 35 mm rather than 20 mm slows blockages compared with a smaller diameter without altering the thermal transfer;

– a slurry speed in the order of 2 $m.s^{-1}$ ensures that any crystals deposited will be reintroduced without causing any erosion on the tube wall.

4.10.7. *Checking the crystallization*

Many instruments are required in order to follow crystallizer operations. We will briefly review these instruments:

– vapor temperature is easy to measure by means of a thermowell situated in the vapor space. This temperature is identical to that of the magma. For instance, with sugar, this temperature is between 70 and 85°C. An increase of 10°C multiplies the growth rate of the crystals by 1.7;

– magma level is determined by the pressure on a membrane situated at a location without any crystal deposits (incrustation);

– the crystals' dispersion flowrate is obtained by means of an electromagnetic flow meter rather than by the pressure difference around a diaphragm, as with clear liquids;

– the dry material content of the mother liquor is taken by a refractometer (measured from the refraction index). This continuous measurement is independent of the presence of gas bubbles. Microwave dephasing is also a possibility;

– crystal content is taken by γ-ray absorption, supplied by a Cesium-137 radioactive source. Protection for personnel should be supplied for this. Ultrasound diffraction is also a possibility;

– granulometry of crystals dispersed in a liquid is obtained continuously by diffraction of a light source by the crystals. The dispersion must flow through the tube at a constant speed. The device records several successive images that are analyzed by scanning, in which the clear patches correspond to crystals. Granulometry frequency or cumulative frequency is established by a software system. This system is used for sugar.

4.10.8. *Surface pollution*

1) Splashes in the vaporization space

The vapor velocity must be limited in the gas part of vacuum crystallizers, so that the walls are not covered with droplets, which would leave a solid deposit on vaporization.

According to Mersmann [MER 88], the vapor velocity must be such that:

$$V_B < 0,1\left[\frac{\gamma(\rho_L - \rho_B)g}{\rho_B^2}\right]^{1/4} \qquad (\text{m.s}^{-1})$$

γ : surface tension of the liquid (N.m^{-1})

ρ_L , ρ_B : density of the liquid and the condensation (kg.m^{-3})

g: acceleration due to gravity (9.81 m.s^{-2}).

2) Encrustation of surfaces and thermal transfer

Encrustation is detrimental to thermal transfer.

The thermal flow density is:

$$q = \alpha_{sus}(t_{sus} - t_{surf}) = K(t_{sus} - t_f)$$

where t_{sus}, t_{surf}, t_{pf} and t_f are, respectively, the temperatures of the suspension, the wall surface (encrusted or not), the metal lining for the thermal fluid and, finally, the thermal fluid (the liquid that heats or cools, gas which condenses, liquid which vaporizes). The thermal flux density is:

By adding the thermal resistances:

$$\frac{1}{K} = \frac{1}{\alpha_{sus}} + \frac{e_M}{\lambda_M} + \frac{e_{en}}{\lambda_{en}} + \frac{i}{\alpha_f}$$

e_M : thickness of the metal lining: 0.001–0.002 m

λ_M: metal conductivity ($\text{W.m}^{-1}.\text{°C}^{-1}$) (15 for stainless steel and 50 for mild steel)

e_{en} : encrustation thickness (m) (zero initially)

λ_{en} : conductivity of deposit ($\text{W.m}^{-1}.\text{°C}^{-1}$) (from 1 to 5)

α_{sus} and α_f are the forced convection coefficients for the suspension and for the thermal fluid.

If the solution includes a solute that can crystallize on the wall, it is necessary that:

$$\left|t_{sus} - t_{surf}\right| = \frac{K}{\alpha_{sus}}\left|t_{sus} - t_f\right| < \Delta t_{lim} \qquad \text{with} \qquad 2°C < \Delta t_{lim} < 5°C$$

Wall roughness has a detrimental effect for two reasons:

– protuberances hook particles from the suspension and can thus hold them to the wall;

– the wall surface area is multiplied by the roughness compared with a smooth surface, which facilitates nucleation (see section 4.1.2.).

All deposits are subject to re-entrainment, if the suspension velocity is sufficient.

There are three types of deposits:

1) thickness e_{en} increases in a linear way according to the time for hard and resistant deposits;

2) if the deposit is less resistant, the fluid can at least partially drag it. According to time, thickness e_{en} follows a law of logarithmic aspect;

3) re-entrainment is equal to the deposit, or when thickness e_{en} is such that the global transfer coefficient has decreased sufficiently, encrustation ceases and e_{en} moves towards an asymptomatic level.

Note that crystal deposit on the thermal transfer surfaces can be significant, particularly for crystallizers; this deposit is known as a "crust".

Often, encrustation only begins after a latency period during which the particles are progressively snared by the wall, which occurs more slowly if the suspension velocity is high (the recommended velocity is between 1.5 $m.s^{-1}$ and 2 $m.s^{-1}$)

4.10.9. *Precipitation by chemical reaction*

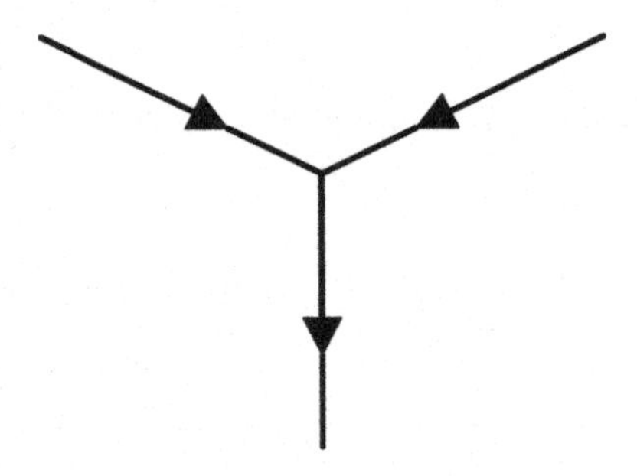

Figure 4.13. *Strong nucleation (fine particles). Mix in Y*

If we mix the two reactives continuously in Y, supersaturation is maximum, as the reactives are not diluted. This is the opposite of what happens in a reactor, where the arriving reactives are diluted in the liquid present in the device in which crystals are already large in size.

Y mixing of the product produces crystals of the order of a micrometer, or even nanometers. In an agitated recipient, if we introduce both reactives immediately on discharge from the impeller, we approach the Y procedure.

Inversely, an active circulation in the recipient benefits dilution and slows nucleation. However, too energetic an agitation generates attrition.

In the volume unit, the mass balance is written as:

$$(1-\varphi_T)\frac{dc}{dt} = -\frac{1}{2}G\,a\,\rho_c$$

But:

$$a = \frac{6\varphi_T}{L_{32}} \quad \text{and} \quad G = 2k\frac{\Delta c}{\rho_c} \quad \text{and} \quad \frac{dc}{dt} = \frac{d(\Delta c)}{dt}$$

k: transport coefficient $(m.s^{-1})$

c: concentration $(kg.m^{-3})$;

hence, after integration:

$$\mathrm{Ln}\left(\frac{\Delta c}{\Delta c_0}\right)=\frac{-6\varphi_T kt}{(1-\varphi_T)L_{32}}$$

where Δc_o is the initial supersaturation (maximum).

We observe that the decrease in supersaturation Δc is faster if the φ_T volume fraction of the solid is large, and if the mean harmonic dimension of the particles is small.

4.11. Definition of a crystallization micropilot

4.11.1. *Use of the pilot*

First, we choose the type of crystallizer that we believe best suited to ensure that the crystal granulometry obtained is suited to our requirements going forward:

– liquid–solid separation and drying;

– ease of use for clients.

We must then ensure the merits of this choice by means of a pilot, which must be:

– cheap, so small;

– easily extrapolated according to clear laws;

– fitted with pipes that do not constantly become blocked.

The pilot must allow us to study the influence of the kinetic parameters on granulometry (and crystallization yield if we proceed by cooling). These parameters are as follows:

– residence time of the slurry in the crystallizer;

– temperatures required by the slurry at various places in the installation;

– slurry crystal content;

– recirculation flow through the exchanger.

In addition:

– choice of temperature, to avoid crusting the exchanger while remaining economical;

– sensitivity of crystals to stress imposed by pumps and agitators (mechanical fragility).

If the pilot study is performed correctly, the industrial exchanger will not become blocked and *a fortiori*, neither will the tubes. We will consider several principles to adhere to for a correct pipe design.

4.11.2. *Definition of pilot exchanger tubes: velocity in the tubes*

This device must be carefully designed to lend itself to the study of tube crusting, to ensure that the extrapolation does not lead to any blockage in the tubes of the industrial exchanger.

To this end, it is necessary that:

– the entry and exit temperatures for the process fluid are the same as that in the industrial installation;

– the temperature of the internal wall is the same.

4.11.3. *Velocity in the pilot tubes*

In an industrial exchanger treating a crystal suspension, we use tubes with an interior diameter of 35 mm. On the other hand, in a pilot, we use 1 or 2 tubes with an interior diameter of 20 mm.

To ensure that the temperature of the internal wall is the same in both installations, and accounting for the fact that the conditions outside the tubes are similar, it should suffice that the partial thermal transfer coefficient inside the tubes is the same. Of course, the entry and exit temperatures of the exchangers are the same in both installations.

If we apply the turbulent regime, coefficient α_i is given by:

$$\alpha_i = 0.023 \frac{\lambda_F}{d_i} (\mathrm{Re})^{0.8} (\mathrm{Pr})^{0.33}$$

λ_F : fluid thermal conductivity (W.m^{-1}.°C^{-1})

V: tube velocity of process fluid (m.s^{-1})

d_i: tube interior diameter (m).

$$Re = \frac{V d_i \rho}{\mu}$$

So, α_i is proportional to: $\left[V^{0.8}.\, d_i^{-0.2}\right]$

Or rather:

$$\frac{\alpha_{i2}}{\alpha_{i1}} = 1 = \left(\frac{V_2}{V_1}\right)^{0.8} \times \left(\frac{d_{i1}}{d_{i2}}\right)^{0.2}$$

Index 1 characterizes the pilot and index 2 the industrial installation.

Therefore:

$$\frac{V_2}{V_1} = \left(\frac{d_{i2}}{d_{i1}}\right)^{1/4}$$

Accordingly, if:

$d_{i1} = 0.020$ $\quad d_{i2} = 0.035$ $\quad$ and $V_2 = 2$ m.s^{-1}

$$V_1 = 2 \times \left(\frac{20}{35}\right)^{0.25} = 1.73\,\text{m.s}^{-1}$$

4.11.4. *Friction on the tube wall*

The pressure drop in a tube is expressed by:

$$\Delta P = \frac{2\pi r L \tau}{\pi r^2} = f\frac{1}{2}\rho V^2 \frac{L}{D} \quad \text{with} \quad D = 2\,r$$

Hence, the friction stress:

$$\tau = \frac{1}{8} f \rho V^2$$

Moreover, for two different tubes:

$$\frac{\tau_2}{\tau_1} = \frac{f_2 V_2^2}{f_1 V_1^2}$$

For an exchanger tube, the friction coefficient f is given by:

$$f = 0.0014 + \frac{0.125}{Re^{0.32}} \quad \text{([KER 50, p. 53])}$$

With:

$$Re = \frac{VD\rho}{\mu}$$

EXAMPLE.–

1) Industrial tubes:

$V_2 = 2\ m.s^{-1}$ $\qquad$ $D_2 = 0.035\ m$

And if we transport water:

$\mu = 10^{-3}\ Pa.s$ $\qquad$ $\rho = 1{,}000\ kg.m^{-3}$

$$Re = \frac{2 \times 0.035 \times 1{,}000}{10^{-3}} = 70{,}000$$

$$f = 0.0014 + \frac{0.125}{(70{,}000)^{0.32}} = 0.00492$$

2) Pilot tubes:

$V_1 = 1.73\ m.s^{-1}$ $\qquad$ $D_2 = 0.020\ m$

$$Re = \frac{1.73 \times 0.02 \times 1{,}000}{10^{-3}} = 34{,}600$$

$$f = 0.0014 + \frac{0.125}{(34{,}600)^{0.32}} = 0.00581$$

3) Friction stress:

$$\frac{\tau_1}{\tau_2} = \frac{0.00581}{0.00492}\left(\frac{1.73}{2}\right)^2 = 0.88$$

The friction on the pilot exchanger tube wall is weaker, which will lead to the apparition of crusting if it can form. Inversely, if the pilot tubes remain clean, we can be certain that the tubes of the industrial exchanger will also remain clean.

4.11.5. *How should the blockages in the pilot be avoided?*

The operation of a pilot is inevitably punctuated by stoppage periods during which crystals can decant rapidly wherever they may be. In the presence of a saturated mother liquor, the sediment can, in certain cases, become resistant in the space of a few minutes. Thus, horizontal pipes are to be avoided. Only vertical pipes or pipes inclined at 45° should be used. Moreover, the exchanger itself should be of the vertical variety.

Valves must, as far as possible, be direct passage valves and must always be placed on the high points of pipes. In particular, control valves can close in a brief instant, which can be enough to cause a blockage if the solid accumulates around the valve.

Valves should be of a nominal diameter equal to (and not less than) that of the pipe.

During a stoppage, slurry decants to the low points of the pipes and the solid accumulates there. Accordingly, at each of these points, a cleaning liquor (often hot water) injection pipe should be fitted. The exit of this liquor is at a high point.

We will not return to slurry velocity in the exchanger tubes. However, the velocity in the tubes must be between 1.5 and 3 $m.s^{-1}$. Indeed, we can accept a higher velocity in a pilot, even if this entails the erosion of the pipe wall.

These walls are not made to endure for more than several months. A well-planned pilot test should last for 3 weeks.

4.11.6. *Conclusion*

If we set the interior diameter and the number of tubes in the pilot exchanger, the rest should follow naturally, particularly the volume of the crystallizer.

Of course, we should retain the following parameters in the industrial installation:

– the crystal content in the slurry;

– the residence time in the crystallizer;

– the renewal velocity of the crystallizer volume by flow through the exchanger;

– the temperature at various points in the installation.

4.12. Crystallization at high temperature

4.12.1. *General aim [BYR 02]*

In general, the aim is to obtain the crystals of inorganic compounds with a high melting point.

For jewelry, "large crystals", whose size is in the order of a centimeter, are sought. This result can be reached by whichever one of the following three methods is appropriate:

– the Verneuil method;

– the Czochralski method;

– the thermal gradient method.

For electronics, we manufacture semi-conductor "films" that are monocrystals in a thin layer.

Uses can also be found for crystals sized in the order of a millimeter.

4.12.2. *The Verneuil method*

The powdered material is treated by a welding torch that turns it into fine droplets, which are then deposited on the crystal by forming a small pool. This pool feeds the crystal.

4.12.3. *The Czochralski method*

A crystal that is initially small in size is brought into contact with the surface of the product in a melted state, which then feeds the crystal. The growing crystal is moved by rotating its vertical axis, and in addition, rises slowly as it develops. Ultimately, the crystal has a cylindrical form that can be cut into fine slices (silicon, semi-conductors). This process requires considerable expertise.

4.12.4. *The thermal gradient method*

The procedure is performed in a small, vertically elongated autoclave that can operate under pressures of 100–2,000 bar and temperatures of 300–1,000°C according to requirements.

The highest temperature is in the lower part of the autoclave, which is where the raw material is to be dissolved. Crystallization occurs at the other cooler, end on a seed placed beforehand. Transport is performed by convection movements of the solvent.

If the solvent is an aqueous solution, we refer to this as the hydrothermal method. The use of water allows us to operate at below the melting temperature in the space that we want to crystallize.

However, the solvent can be a metallic oxide and subsequently water soluble, which allows for recovery of the crystal, which is not water soluble.

The silicon carbide does not melt, but sublimes. A mixture of silica and coke fuse at 2,500°C under an argon atmosphere. The result is a SiC vapor, which is condensed into plates of 1 cm^2 in surface on a cooler wall. This is the Lely process.

4.12.5. *Film production*

The material is typically brought into vapor phase on a horizontal plane substrate. If the material of the substrate is different from that of the film in production, we refer to it as epitaxy. This can be used for:

– a molecular cluster in a high vacuum (10^{-9} to 10^{-8}Pa). This technique is used for semi-conductors;

– a heated gas jet. For example, to make a diamond film, the gas is a hydrocarbon (CH_4) heated either by an incandescent filament or a plasma torch. The substrate can be silica;

– the sandwich method. The raw material “source” is planar and placed at a distance in the order of a millimeter from the substrate, which is also planar. This method is used for silicon carbide (which sublimes easily, as we have seen).

4.12.6. *Crystals in the order of a millimeter*

The solvent can be a metallic oxide with low melting point and be water soluble. We proceed in the normal way with one of the following two methods:

– slow cooling;

– solvent vaporization.

Batch crystallization theory applies to this method.

4.13. Calculation for continuous crystallizers

4.13.1. *Cooling*

We can deduce the composition X_s of the exit liquor by:

– knowing the composition X_e and the temperature T_e of the supply, that is at the entry, and;

– providing ourselves with the mass of crystals to be produced W_E (X_e-X_s).

The temperature T* of this liquor must correspond to the exiting saturation equilibrium.

Thus, we know the entry temperature T_e and exit temperature T* of the mass. We can deduce the thermal cooling power to be provided Q:

$$R = (T_e - T^*)(W_E C_E + W_E X_e C_S) + W_E (X_e - X_s) \Lambda_c$$

If the difference in concentration between equilibrium X* and the level of real solute on exit X_s is assumed to be zero, experience shows that this hypothesis is typically verified.

Q: thermal power (W)

T_e: entry temperature of the mother liquor (degrees Celsius)

T^*: exit temperature of the slurry (degrees Celsius)

W_E: flow of water accompanying the product to be crystallized ($kg.s^{-1}$)

X_e: level of solute in the mother liquor on entry ($kg.kg^{-1}$)

X_S: level of solute on exit ($kg.kg^{-1}$)

C_E: thermal capacity of water ($J.kg^{-1}.°C^{-1}$)

C_S: thermal capacity of the product to be crystallized ($J.kg^{-1}.°C^{-1}$)

Λ_C : crystallization heat of the product ($J.kg^{-1}$).

4.13.2. *Solvent vaporization*

We know the composition and temperature of the supply, and provide ourselves with the crystal mass to be produced. The operating pressure of the crystallizer is given.

We can thus deduce the saturation temperature T^* of the mother liquor for the entry concentration.

During vaporization, concentration does not change as the slurry temperature remains constant and equal to T^*, since the pressure remains constant and equal to steam pressure π at T_E. The difference between T^* and T_E rarely exceeds 5°C, as a general rule. This difference is the boiling delay.

Thus, the thermal heating power Q is written as:

$$Q = \left(T^* - T_e\right)\left(W_{Ee}C_E + W_{Ee}XC_s\right) + \Lambda_E\left(W_{Ee} - W_{Es}\right)$$

This power is the total of the heat corresponding to the preheating and the water vaporization heat.

Q: thermal power (W)

T_e: entry temperature of the mother liquor (°C)

T^*: slurry exit temperature (°C)

W_{Ee}: water flow on entry ($kg.s^{-1}$)

W_{Es}: water flow on exit ($kg.s^{-1}$)

C_E: thermal capacity of water ($J.kg^{-1}.°C^{-1}$)

C_S: thermal capacity of the product to crystallize ($J.kg^{-1}.°C^{-1}$)

Λ_E : vaporization heat of water ($J.kg^{-1}$)

X: mass ratio of the product ($kg.kg^{-1}$).

4.13.3. *Establishing the thermal transfer parameters*

There are two parameters as follows:

– global transfer coefficient U ($W.m^{-2}.°C^{-1}$);

– the surface through which the transfer occurs (m^2).

Coefficient U can be calculated with the rules of a thermal calculation (see [DUR 16]). It can also be established by tests:

1) If the solvent is water, we can:

– circulate water continuously in the crystallizer and use the classic calculation for thermal exchangers;

– treat a given mass of water filling the crystallizer and apply the laws of the thermal transfer;

2) if the solvent is not water, we can try a mass of pure solvent filling the crystallizer;

3) there are effectively only two types of thermal source:

– heating steam (solvent vaporization);

– brine (typically a $CaCl_2$ solution in water) or simply cooling water.

Recall that the heat transferred over time $\Delta\tau$ through the surface element dS is:

$$dq = UdS(T_S - T_B)\Delta\tau$$

T_S: temperature of the heat (or cold) source (°C)

T_B: slurry temperature in the crystallizer (°C).

APPENDICES

Appendix 1

Numerical Integration: The Fourth-order Runge–Kutta Method

The aim is to integrate the differential equation:

$$\frac{dx}{d\tau} = F(x, \tau)$$

$$x_{\tau=0} = x_0$$

We write:

$$\tau_{i+1/2} = \tau_i + \frac{\Delta\tau}{2}$$

$$x_{i+1/2}^{(1)} = x_i + \frac{\Delta\tau}{2} F(x_i, \tau_i)$$

$$x_{i+1/2}^{(2)} = x_i + \frac{\Delta\tau}{2} F(x_{i+1/2}^{(1)}, \tau_{i+1/2})$$

$$x_{i+1}^{(1)} = x_i + \Delta\tau\, F(x_{i+1/2}^{(2)}, \tau_{i+1/2})$$

Hence:

$$x_{i+1} = x_i + \frac{\Delta\tau}{6}\left[F(x_i, \tau_i) + 2F(x^{(1)}_{i+1/2}, \tau_{i+1/2})\right.$$

$$\left. +2F(x^{(2)}_{i+1/2}, \tau_{i+1/2}) + F(x^{(1)}_{i+1}, \tau_{i+1})\right]$$

We can generalize the method to a system of n differential equations of the first order applying n variables x_j (j of 1 to n). The independent variable is x_0.

$$\frac{dx_j}{dx_0} = F_j(x_0, x_1, \ldots, x_j, \ldots, x_n)$$

We write:

$$x_{0,i+1} = x_{0,i} + \Delta x_0 \qquad \text{and} \quad x_{0,i+1/2} = x_{0,i} + \frac{\Delta x_0}{2}$$

$$x^{(1)}_{j,i+1/2} = x_{j,i} + \frac{\Delta x_0}{2} F_j(x_{0,1}, \ldots, x_{j,i}, \ldots x_{n,i})$$

$$x^{(2)}_{j,i+1/2} = x_{j,i} + \frac{\Delta x_0}{2} F_j\left(x_{0,i+1/2}, \ldots, x^{(1)}_{j,i+1/2}, \ldots, x^{(1)}_{n,i+1/2}\right)$$

$$x^{(1)}_{j,i+1} = x_{j,i} + \Delta x_0 F\left(x_{0,i+1/2}, \ldots, x^{(2)}_{j,i+1/2}, \ldots, x^{(2)}_{n,i+1/2}\right)$$

And finally:

$$x_{j,i+1} = x_{j,i} + \frac{\Delta x_0}{6}\left[F_j(x_{0,i}, \ldots x_{j,i}, \ldots, x_{n,i}) + 2F_j\left(x_{0,i+1/2}, \ldots x^{(1)}_{j,1+1/2}, \ldots x^{(1)}_{n,i+1/2}\right)\right.$$

$$\left. +2F_j\left(x_{0,i+1/2}, \ldots x^{(2)}_{j,i+1/2}, \ldots x^{(2)}_{n,i+1/2}\right) + F_j\left(x_{0,i+1}, \ldots x^{(1)}_{j,i+1}, \ldots x^{(1)}_{n,i+1}\right)\right]$$

Appendix 2

Resolution of Equations of the Third and Fourth Degree Searching for Dimensionless Groups

A2.1. Third-degree equation

Consider the cubic equation:

$$x^3 + \alpha x^2 + \beta x + \gamma = 0$$

We write:

$$p = \beta - \frac{\alpha^2}{3}$$

$$q = \gamma - \frac{\alpha\beta}{3} + \frac{2\alpha^3}{27}$$

Hence:

1) If $\frac{q^2}{4} + \frac{p^3}{27} > 0$

$$x = -\frac{\alpha}{3} + \left[\frac{-q}{2} + \sqrt{\frac{q^2}{4} + \frac{p^3}{27}}\right]^{1/3} + \left[\frac{-q}{2} - \sqrt{\frac{q^2}{4} + \frac{p^3}{27}}\right]^{1/3}$$

2) If $\frac{q^2}{4}+\frac{p^3}{27}<0$

$$y_1 = 2\sqrt{\frac{-p}{3}}\cos\frac{a}{3};\ y_2 = 2\sqrt{\frac{-p}{3}}\cos\left(\frac{a}{3}+\frac{2\pi}{3}\right);\ y_3 = 2\sqrt{\frac{-p}{3}}\cos\left(\frac{a}{3}+\frac{4\pi}{3}\right)$$

With:

$$a = \cos^{-1}\frac{3q}{p\rho},\quad \rho^2 = \frac{-4p}{3};\quad x_i = y_i - \frac{\alpha}{3}$$

A2.2. Fourth-degree equation

The most general form of a fourth-degree equation is:

$$x^4 + d_3x^3 + d_2x^2 + d_1x + d_0 = 0$$

We write:

$$\delta = d_3/2 \quad \alpha = 2A - \delta^2$$

$$A = d_2 - \delta^2 \qquad \beta = A^2 + 2B\delta - 4d_0$$

$$B = d_1 - A\delta \qquad \gamma = -B^2$$

Now consider the cubic equation:

$$\xi^3 + \alpha\xi^2 + \beta\xi + \gamma = 0$$

As the expression to the left of this equality is negative for $\xi = 0$, the equation certainly has a positive root ξ.

We can now write:

$$\delta = d_3/2 \quad \varepsilon = \frac{1}{2}(A+\xi)$$

$$d = \sqrt{\xi} \qquad e = \frac{d}{2}\left(\delta - \frac{B}{\xi}\right)$$

In addition, the solutions of the fourth-degree equation are those of both second-degree equations:

$$x^2 + (\delta + d)x + (\varepsilon + e) = 0$$

$$x^2 + (\delta - d)x + (\varepsilon - e) = 0$$

A2.3. Searching for dimensionless groups expressing a physical law

Now consider a physical law applying n expressible physical quantities using k fundamental magnitudes. Thus, the physical quantity velocity is the quotient of length by time.

Buckingham shows that the number of adimensional groups is:

$$i = n - k$$

On page 349 of [BUC 14], he gives an example of the calculation of i adimensional groups.

However, the author shows that the number of possible adimensional groups (each including i groups) is equal to i. We must then refer to the physics of the phenomena in order to choose a system to apply.

From page 356 of [BUC 14], the author begins to consider units of electromagnetism. However, these considerations are outdated today.

On searching for dimensionless groups, we can consult: Buckingham [BUC 14] and [BUC 15].

Bibliography

[AGL 79] AGLER A.T., LIFSON S., DAUBER P., "Consistent force field studies of intermolecular forces in hydrogen-bonded crystals. 2. A benchmark for the objective comparison of alternative force fields", *Journal of the American Chemical Society*, vol. 101, pp. 5122–5130, 29 August 1979.

[AUS 81] AUSTMEYER K., Untersuchungen zum Wärme und Stoffübergang im Anfangsstadium der Verdampfungskristallisation der Saccharose, Dissertation, Technical University, Braunschweig, 1981.

[BEN 68] BENNEMA P., "Surface diffusion and the growth of sucrose crystals", *Journal of Crystal Growth*, vols. 3–4, p. 331, 1968.

[BES 69] BESSET R., "Continuous sugar crystallization: a chemical engineer's viewpoint", *Chemical Engineering Progress Symposium Series*, vol. 65, p. 34, 1969.

[BRA 66] BRAVAIS A., *Etudes cristallographiques*, Editions Gauthier-Villars, Paris, 1866.

[BRO 92] BROWN D.J., ALEXANDER K., "Rates of nucleation in the crystallization of sucrose", *Journal of Crystal Growth*, vol. 118, p. 464, 1992.

[BRU 96] BRUHNS M., *Fliessverhalten von Zuckerkristallsuspensiaren und Wärmeübergang bei der Zucker Verdampfungskristallisation bei kleinen Temperaturgefällen*, Verlag Dr Albert Bartens, Berlin, 1996.

[BUB 84] BUBNIK Z., KADLEC P., "Geschvindigkeit der Saccharosekristallisation", *Internationaler Kongress CHISA in Prague Zuckerindustrie*, vol. 109, no. 12, p. 1117, 1984.

[BUB 95] BUBNICK Z., KADLEC P., URBAN D. *et al.*, *Sugar Technologists Manual*, Verlag Dr. Albert Bartens, Berlin, 1995.

[BUC 14] BUCKINGHAM E., "On physically similar systems. Illustrations of the use of dimensional equations", *Physical Review*, vol. 4, no. 4, pp. 345–376, 1914.

[BUC 15] BUCKINGHAM E., "The principle of similitude", *Nature*, vol. 96, pp. 396–397, 1915.

[BUC 47] BUCKINGHAM R.A., "Tables of second virial and low-pressure Joule–Thomson coefficient for intermolecular potentials with exponential repulsion", *Proceedings of the Royal Society*, vol. A189, p. 118, 1947.

[BUR 51] BURTON W.K., CABRERA N., FRANK F.C., "The growth of crystals and the equilibrium structure of their surfaces", *Philosophical Transactions of the Royal Society*, vol. A243, p. 299, 1951.

[BYR 02] BYRAPPA K., OHACHI T., *Crystal Growth Technology*, Springer, New York, 2002.

[CHA 84] CHANG Y.C., MYERSON A.S., "Diffusion coefficients in supersaturated solutions", in JANCIC S.J., DE JONG E.J. (eds), *Industrial Crystallisation*, Elsevier, Amsterdam, 1984.

[CHE 61] CHERNOV A.A., "The spiral growth of crystals", *Uspekhi Fizicheskikh Nauk*, vol. 73, p. 277, 1961.

[COR 98] CORIELL S.R., CHERNOV A.A., MURRAY B.T. *et al.*, "Step bunching: generalized kinetics", *Journal of Crystal Growth*, vol. 183, p. 669, 1998.

[CUR 85] CURIE P., "Sur la formation des cristaux et sur les constantes capillaires de leurs diverses faces", *Bulletin de la Société minéralogique de France*, vol. 8, p. 145, 1885.

[DAV 85] DAVIDSON J.F., CLIFT R., HARRISON D., *Fluidization*, 2nd ed., Academic Press, 1985.

[DEV 83] DE VRIES G.H., "New method for the continuous crystallisation of sugar", *Sugar Technology Review*, vol. 10, p. 3, 1983.

[DIA 73] DIALER K., KÜBNER K., "Oberflächenentfaltung und Energieaufname bei der Schwingmalhung von Kristallzucker zum Einfluβ der Bindungsverhältnisse auf die Feinstzerkleinerung", *Kolloid-Zeitschrift und Zeitschrift für Polymere*, vol. 251, pp. 710–715, 1973.

[DIR 91] DIRKSEN J.A., RING T.A., "Fundamentals of crystallisation: kinetic effects on particle size distributions and morphology", *Chemical Engineering Science*, vol. 46, p. 2389, 1991.

[DOW 80] DOWTY E., "Computing and drawing crystal shapes", *American Mineralogist*, vol. 65, p. 465, 1980.

[DUD 86] DUD'A R., REJL L., *La grande encyclopédie des minéraux*, Editions Gründ, 1986.

[DUR 16] DUROUDIER J.-P., *Heat Transfer in the Chemical, Food and Pharmaceutical Industries*, ISTE, London and Elsevier, Oxford, 2016.

[FRI 07] FRIEDEL M.G., "Etudes sur la loi de Bravais", *Bulletin de la Societé française de Mineralogie*, vol. 9, p. 326, 1907.

[GAH 97] GAHN C., Die Festigkeit von Kristallen und ihr Einfluβ auf die Kinetik in Suspensions kristallisatoren, Thesis, Technical University, München, 1997.

[GAV 94] GAVEZZOTTI A., "Are crystal structures predictable?", *Accounts of Chemical Research*, vol. 27, p. 309, 1994.

[GRI 98] GRIMBERGEN R.F.P., MEEKES H., BENNEMA P. *et al.*, "On the prediction of crystal morphology, Part I, The Hartmann–Perdok theory revisited", *Acta Crystallographica*, vol. A54, p. 491, 1998.

[GRI 99] GRIMBERGEN R.F.P., BENNEMA P., MEEKES H., "On the prediction of crystal morphology, Part III, Equilibrium and growth behaviour of crystal faces containing multiple connected nets", *Acta Crystallographica*, vol. A55, p. 84, 1999.

[HAG 79] HAGLER A.T., LIFSON S., DAUBER P., "Consistent force field studies of intermolecular faces in hydrogen – bonded crystals. Part II", *Journal of the American Chemical Society*, vol. 101, p. 5122, 1979.

[HAR 55a] HARTMANN P., PERDOK W.G., "On the relation between structure and morphology of crystals. Part I", *Acta Crystallographica*, vol. 8, p. 49, 1955.

[HAR 55b] HARTMANN P., PERDOK W.G., "On the relation between structure and morphology of crystals. Part II", *Acta Crystallographica*, vol. 8, p. 521, 1955.

[HAR 55c] HARTMANN P., PERDOK W.G., "On the relation between structure and morphology of crystals. Part III", *Acta Crystallographica*, vol. 8, p. 525, 1955.

[HAR 80] HARTMANN P., BENNEMA P., "The attachment energy as a habit controlling factor", *Journal of Crystal Growth*, vol. 49, p. 145, 1980.

[HAR 96] HARTEL R.W., "Controlling crystallization in foods", *Journal of the American Chemical Society*, pp. 172–177, 1996.

[JAN 84] JANCIC S.J., GROOTSCHOLTEN P.A.M., *Industrial Crystallization*, Delft University Press, 1984.

[JON 74a] JONES A.G., "Optimal operation of a batch cooling crystallizer", *Chemical Engineering Science*, vol. 29, p. 1075, 1974.

[JON 74b] JONES A.G., MULLIN J.W., "Programmed cooling crystallization of potassium sulphate solutions", *Chemical Engineering Science*, vol. 29, p. 105, 1974.

[KER 50] KERN D.R., *Process Heat Transfer*, McGraw Hill, New York, 1950.

[KUR 96] KURIHARA K., MIYASHITA S., SAZAKI G. *et al.*, "Interferometric study on the crystal growth of tetragonal lyzozyme crystal", *Journal of Crystal Growth*, vol. 166, p. 904, 1996.

[LEC 68] LECI C.L., MULLIN J.W., "Refractive index measurements in liquids rendered opaque by the presence of suspended solids", *Chemistry and Industry*, p. 1517, 1968.

[LEW 74a] LEWIS B., "The growth of crystals at low supersaturation. Part I. Theory", *Journal of Crystal Growth*, vol. 21, p. 29, 1974.

[LEW 74b] LEWIS B., "The growth of crystals at low supersaturation. Part II. Comparaison with experiment", *Journal of Crystal Growth*, vol. 21, p. 40, 1974.

[LIA 87] LIANG B.M., HARTEL R.W., BERGLUND K.A., "Contact nucleation in sucrose crystallization", *Chemical Engineering Science*, vol. 42, no. 11, p. 2723, 1987.

[LIF 79] LIFSON S., HAGLER A.T., DUABER P., "Consistent force field studies of intermolecular forces in hydrogen – bonded crystals Part I", *Journal of the American Chemical Society*, vol. 101, p. 5111, 1979.

[LIU 96] LIU X.-Y., BENNAMA P., "Theoretical considerations of the growth morphology of crystals", *Physical Review*, vol. B53, p. 2314, 1996.

[MAL 89] MALKIN A.I., CHENOV A.A., ALEXEEV I.V., "Growth of dipyramidal face of dislocation-free ADP crystals: free energy of steps", *Journal of Crystal Growth*, vol. 97, p. 765, 1989.

[MAR 84] MARQUSEE J.A., ROSS J., "Theory of Ostwald ripening: competitive growth and its dependence on volume fraction", *Journal of Chemical Physics*, vol. 80, no. 1, pp. 536–543, 1984.

[MAU 85] MAURANDI V., MANTOVANI G., VACCARI G., "Sucrose crystallisation at low supersaturation in impure beet syrups and pure solutions", *Zuckerindustrie*, vol. 110, p. 1096, 1985.

[MEE 98] MEEKES H., BENNEMA P., GRIMBERGEN R.F.P., "On the prediction of crystal morphology. Part II. Symmetry roughening of pairs of connected nets", *Acta Crystallographica*, vol. A55, p. 501, 1998.

[MER 88] MERSMANN A., "Design of cystallizers", *Chemical Engineering and Processing*, vol. 23, p. 213, 1988.

[MER 01] MERSMANN A., *Crystallization Technology Handbook,* Marcel Dekker, New York, 2001.

[MIL 47] MILLER P., SAEMAN W.C., "Continuous vacuum crystallization of ammonium nitrate", *Chemical Engineering Progress*, vol. 43, no. 12, p. 667, 1947.

[MIL 69] MILAZZO G., *Electrochimie*, Editions Dunod, Paris, 1969.

[MOM 74] MOMANY F.A., CARRUTHERS L.M., MC GUIRE R.F. *et al.*, "Intermolecular potentials from crystals data – Part III Détermination of empirical potentials and application to the packing configurations and lattice energy in crystals of hydrocarbons, carboxylic acids, amines et amides", *The Journal of Physical Chemistry*, vol. 78, p. 1595, 1974.

[MUL 72] MULLIN J.W., *Crystallisation*, 2nd ed., Butterworths, 1972.

[MUT 01] MUTAFTSCHIEV B., *The Atomistic Nature of Crystal Growth*, Springer, New York, 2001.

[NAG 75] NAGATA S., *Mixing*, Halstead Press, Sydney, 1975.

[NEM 83] NEMETHY G., POTTLE M.S., SCHERAGA H.A., "Energy parameters in polypeptides. 9. Updating of geometrical parameters, nonbonded interactions and hydrogen bond interactions for the naturally occurring amino acids", *Journal of Physical Chemistry*, vol. 87, p. 1883, 1983.

[NIE 71] NIELSEN A.E., SÖHNEL O., "Interfacial tensions electrolyte crystal–aqueous solution from nucleation data", *Journal of Crystal Growth*, vol. 11, p. 233, 1971.

[NIE 80] NIELSEN A.E., "Transport control in crystal growth from solution", *Croatica Chemica Acta*, vol. 53, p. 255, 1980.

[NOU 85] NOUGIER J.P., *Méthodes de calcul numérique*, Masson, 1985.

[NOV 55] NOVICK A.S., *Crystal Properties via Group Theory*, Cambridge University Press, Cambridge, 1955.

[NYV 85] NYVLT J., SÖHNEL O., MATUCHOVA M. *et al.*, *The Kinetics of Industrial Crystallization*, Elsevier, Amsterdam, 1985.

[OLD 83] OLDSHUE J.Y., *Fluid Mixing Technology*, McGraw Hill, New York, 1983.

[ONS 44] ONSAGER L., "Part I A two-dimensional model with an order–disorder transition", *Physical Review*, vol. 65, p. 117, 1944.

[ORO 49] OROVAN E., "Fracture and strength of solids", *Reports on Progress in Physics*, vol. 12, p. 185, 1949.

[POE 98] POEL P.W., SCHIWECK H.M., SCHWARTZ T.K., *Sugar Technology: beet and cane sugar manufacture*, Verlag Dr Albert Bartens, Berlin, 1998.

[POH 87] POHLISCH R.J., Einfluss von mechanischer Beeinflussung und Abrieb auf die Korngrössenverteilung, Thesis, Technical University, München, 1987.

[QUÉ 88] QUÉRÉ Y., *Physique des matériaux*, Editions Ellipses, Paris, 1988.

[RAN 71] RANDOLPH A.D., LARSON M.A., *Theory of Particulate Process*, Academic Press, 1971.

[RAT 02] RATKE L., VOORHEES P.W., *Growth and Coarsening*, Springer, New York, 2002.

[ROU 00] ROUSSEAU J.-J., *Cristallographie géométrique et radiocristallographie*, Editions Dunod, 2000.

[SAN 87] SANQUER M., ECOLIVELT C., "Elastic constants in molecular crystals experiments and intermolecular potentials", in LASCOMBE J. (ed.), *Dynamics of Molecular Crystals*, Elsevier, Amsterdam, 1987.

[SCH 83] SCHLIEPHAKE D., EKELHOF B., "Beitrag zur vollständigen Berechnung der Kristallisationsgeschwindigkeit der Saccharose", *Zuckerindustrie*, vol. 108, no. 12, p. 1127, 1983.

[SPI 92] SPIEGEL M.R., *Formules et tables de mathématiques*, McGraw-Hill, New York, 1992.

[STE 90] STEWART J.J.P., "Mopac: a semiempirical molecular orbital program", *Journal of Computer-Aided Molecular Design*, vol. 4, pp. 1–105, 1990.

[TAI 75] TAI C.Y., MC CABE W.L., ROUSSEAU R.W., "Contact nucleation of various crystal types", *AIChE Journal*, vol. 21, no. 2, p. 351, 1975.

[TER 01] TER HORST J.H., GEERTMAN R.M., VAN ROSMALEN G.M., "The effect of solvent on crystal morphology", *Journal of Crystal Growth*, vol. 230, p. 277, 2001.

[ULL 86] ULLRICH M., RATHJEN C., "Fortschritte beim Kristallisieren in Scheckenmasthinen", *Chemie Ingenieur Technik*, vol. 58, no. 7, pp. 590–592, 1986.

[VAN 86] VAN DER EERDEN J.P., "Formation of macrosteps due to time dependant impurity adsorption", *Electrochimica Acta*, vol. 31, p. 1007, 1986.

[VES 99] VESZPREMI T., FEHER M., *Quantum Chemistry Fundamentals to Applications*, Klumer Academic/Plenum Publishing, 1999.

[WAG 62] WAGNEROWSKI K., DABROWSKA D., DABROWSKI C., "Probleme der Melasseerschopfung", *Zeitschrift für die Zuckerindustrie*, vol. 12, p. 664, 1962.

[WAN 92] WANG S., Grossenabhängige Wachstumsdispersion von Abriebsteilchen, Thesis, Technical University, München, 1992.

[WEY 73] WEY J.S., ESTRIN J., "Modeling the batch crystallization process. The ice-brine system", *Industrial & Engineering Chemistry Process Design and Development*, vol. 12, no. 3, p. 236, 1973.

[WIT 92] WITTE G., "Inline Kristallgrössenanalyse", *Chemie Anlagen + Verfahren*, vol. 25, p. 153, 1992.

[WUL 01] WULF G., "Zur Frage der geschwindigkeit des Wachstums und der Auflösung der Kristallfläschen", *Zeitschrift für Kristallographie*, vol. 34, p. 499, 1901.

[YOR 83] YORK P., "Solid-state properties of powders in the formulation and processing of solid dosage forms", *International Journal of Pharmaceutics*, vol. 14, p. 1, 1983.

Index

A, B, C

F, G, H

K, M, N

S, P, V

Printed in the United States
By Bookmasters